AF357332

Boreal Bird Ecology, Management and Conservation

Boreal Bird Ecology, Management and Conservation

Editors

Stacy McNulty
Michale Glennon

MDPI • Basel • Beijing • Wuhan • Barcelona • Belgrade • Manchester • Tokyo • Cluj • Tianjin

Editors
Stacy McNulty
State University of New York College of
Environmental Science and Forestry
USA

Michale Glennon
Paul Smith's College Adirondack
Watershed Institute
USA

Editorial Office
MDPI
St. Alban-Anlage 66
4052 Basel, Switzerland

This is a reprint of articles from the Special Issue published online in the open access journal *Diversity* (ISSN 1424-2818) (available at: http://www.mdpi.com).

For citation purposes, cite each article independently as indicated on the article page online and as indicated below:

LastName, A.A.; LastName, B.B.; LastName, C.C. Article Title. *Journal Name* **Year**, *Volume Number*, Page Range.

ISBN 978-3-0365-2014-8 (Hbk)
ISBN 978-3-0365-2015-5 (PDF)

Cover image courtesy of Shannon Buckley Luepold

Contents

About the Editors

Stacy McNulty is an ecologist and Associate Director of Research and Research Associate for the Adirondack Ecological Center, a biological research station of the State University of New York, College of Environmental Science and Forestry (SUNY-ESF). Stacy has over 20 years of experience in field-based research, project management, and interdisciplinary areas, including organizational governance, natural resource policy, and complex social–ecological systems. Her interests are broad and center on wildlife, forests, waters, and landscapes, studying topics including human–wildlife relationships, phenology, habitat heterogeneity, and the impacts of disturbance and management on northeastern North American ecosystems. Stacy holds degrees from SUNY ESF and SUNY Geneseo.

Michale Glennon is an ecologist and serves as the Science Director for the Adirondack Watershed Institute at Paul Smith's College. She previously spent 15 years as the Director of Science for the Adirondack Program of the Wildlife Conservation Society. She is interested in the effects of land use management on wildlife populations in the Adirondack Park of northern New York State and has 20 years of experience leading ecological research projects focusing on landscape and community ecology using a variety of taxa, including avian, small mammal, and herpetological communities. She has experience with field techniques and analytical methods and publishing scientific findings within various disciplines, including biogeography, conservation, landscape ecology, and resource and recreation management. Her research ranges from issues of residential development to recreation ecology to climate change. Since 2004, Michale has led a long-term project focused on low elevation boreal bird communities in the Adirondacks, changes to those communities over time, and the vulnerability of these species and their peatland habitats in a warming climate. Michale holds degrees from SUNY ESF and Dartmouth College.

Preface to "Boreal Bird Ecology, Management and Conservation"

A recent study in Science reported a 29% decline in the North American bird population over 50 years. The loss of almost 3 billion birds is mirrored in Europe and indicates a widespread ecological crisis that may have resulted from loss and degradation of habitat, especially due to agricultural intensification and urbanization. In the Americas, Partners in Flight lists many Temperate Breeders of High Tri-National Concern (e.g., Bicknell's Thrush, Olive-sided Flycatcher, and Canada Warbler) at risk from habitat loss, contaminants, exotic species, climate change, and threats to their breeding, stopover and wintering grounds. This Special Issue, titled "Boreal Bird Ecology, Management, and Conservation", aims to present the current state of knowledge on boreal bird diversity, ecology, management, and conservation. As an important component of boreal habitats, the taxonomic and functional aspects of the avian community are key to conservation measures and can indicate the effects of anthropogenic activity on boreal habitats worldwide. The original research and observations in this Special Issue will aid policymakers, scientists, and others in understanding and protecting bird species and their boreal habitats.

Stacy McNulty, Michale Glennon
Editors

Editorial

Boreal Bird Ecology, Management and Conservation

Stacy McNulty [1,*], Michale Glennon [2] and Carol Foss [3]

[1] Adirondack Ecological Center, State University of New York College of Environmental Science and Forestry, 6312 Route 28N, Newcomb, NY 12852, USA
[2] Paul Smith's College Adirondack Watershed Institute, Paul Smiths, NY 12970, USA; mglennon@paulsmiths.edu
[3] NH Audubon, Concord, NH 03301, USA; cfoss@nhaudubon.org
[*] Correspondence: smcnulty@esf.edu

Citation: McNulty, S.; Glennon, M.; Foss, C. Boreal Bird Ecology, Management and Conservation. *Diversity* 2021, *13*, 206. https://doi.org/10.3390/d13050206

Received: 6 May 2021
Accepted: 10 May 2021
Published: 13 May 2021

Publisher's Note: MDPI stays neutral with regard to jurisdictional claims in published maps and institutional affiliations.

The circumpolar boreal forest covers approximately 12,000,000 km^2 and is one of the world's most extensive biomes. Despite its geographic extent, the boreal forest supports relatively low avian diversity, with roughly 400 species breeding in the region world-wide [1–3]. Sparse human population, limited transportation infrastructure, short breeding seasons, extensive wetlands, and high densities of blood-sucking insects provide significant challenges for field work in boreal ecosystems, and comprehensive monitoring of most boreal bird populations is largely impractical. Recent evidence of significant declines in North American [4] and European [5] bird populations highlight the importance of understanding the ecology of and threats to boreal species.

This Special Issue of *Diversity*, titled "Boreal Bird Ecology, Management, and Conservation," contributes to the current state of knowledge on boreal bird diversity, ecology, management, and conservation. Nine papers describe bird species' distribution and abundance, biogeography, life history, habitat associations, genetics, population demographics, impact of climate change, land use and planning, and extinction risk across the boreal and temperate/boreal ecotone. As these papers posit, taxonomic and functional aspects of the boreal avian community are key to conservation measures and can indicate effects of anthropogenic activity on boreal habitats worldwide.

In this overview, we highlight new insights and points of synergy between them. We also identify information gaps that hinder population management and the future persistence of these species. The majority of papers in this Special Issue describe boreal birds in a North American context. Several papers highlight issues occurring in the transition zone at the southern edge of the boreal zone in New England and Maritime Canada. Additionally, several papers provide new insight into the ecology of a species of high conservation concern, the Rusty Blackbird (*Euphagus carolinus*). Here, we provide brief overviews of the Special Issue's papers.

Ralston and DeLuca [6] focus on boreal birds at the southern periphery of their range in the northeastern US and southern Canada, and summarize lessons from a decade of research focused on this "battle ground" of environmental change. These authors review and discuss findings of several researchers, including genetic variability at the range margin, impacts of climate change and predicted changes in climate suitability for several species, regional trends for both northeastern and midwestern US boreal bird populations, factors influencing persistence of boreal forest-breeding birds at the southern edge of their distribution, and distributional shifts in response to climate change, community composition, and nest predators. They highlight the unique and specific threats to boreal birds in this geography and the variability in avian responses to changing conditions that pose management challenges for the community as a whole. They also highlight complex shifts in community composition that create a potential for novel ecological interactions and hybridization, and further complicate efforts to find management solutions for declining populations. Open questions remain concerning habitat associations of boreal birds in this region and the degree to which habitat factors associated with stable or increasing populations differ throughout this zone. Additionally, unknown are the long-term consequences of community-level changes resulting in non-analogous communities and new ecological interactions that have evolutionary implications. Continued, expanded, and coordinated monitoring of boreal bird population trends in this geography will be critical to addressing and reversing population declines.

Addressing multiple species of boreal birds in Scandinavia, Virkkala et al. [7] assess the degree to which topographic heterogeneity can increase a landscape's buffering capacity against potential impacts of climate change on boreal birds. They suggest that terrain heterogeneity can buffer effects of climate-induced changes on species and ecological systems by providing local temperature variability over smaller distances compared to flatter, more homogeneous areas. Though most support for this hypothesis comes from modeling studies, the authors investigated temporal changes in bird species densities in protected areas in Finland and their relationship to variation in local temperature, while taking into account broad-scale air temperature (macroclimate) and protected area size. Boreal bird species density was higher in topographically heterogeneous regions, and changes in bird density were smaller in protected areas with higher topographic variation, providing support for a buffering effect stemming from local temperature variation. The topographic variation in their study area was relatively small, suggesting that the buffering impacts of local air temperatures may not require large elevational gradients. These findings provide support for the protection of topographically heterogeneous areas in large tracts where possible, or potentially in a connected network of smaller sites.

In keeping with the theme of heterogeneity and climate change extinction risk, West-wood et al. [8] focused on the Canada Warbler (*Cardellina canadensis*), a species of conservation concern, and where it breeds in Canada and the U.S. The authors evaluated a range of conservation planning scenarios to prioritize areas for permanent land conservation or "responsible forest management" (minimizing species removal during forest harvesting while promoting colonization of regenerated forest). They used Canada Warbler population density, connectivity to protected areas, future climate suitability, anthropogenic disturbance, and recent Canada warbler observations to prioritize potential conservation areas and assessed scenarios with a range of dispersal distances. Large dispersal distances resulted in prioritization of a few large areas, while smaller dispersal distances prioritized smaller, more widely distributed areas. These findings highlight the importance of considering dispersal distance as well as anticipated future habitat conditions in setting priorities for long-term conservation. Empirical data on these aspects of life history may improve similar modeling and planning efforts for other boreal bird species.

Lamarre and Tremblay [9] assessed habitat associations for a declining boreal bird, the American Three-toed Woodpecker, *Picoides dorsalis*. Disturbance from forest management in the eastern range of this woodpecker has caused major losses of old-growth spruce habitat favored by this species. The authors assessed the influence of habitat characteristics at the stand and landscape scales on occupancy of three-toed woodpecker during the breeding season in a heavily managed landscape in eastern Canada. The species exhibited a very low occupancy rate that decreased rapidly with increasing area of recent clearcutting at the stand scale. Occupancy was positively associated with the extent of old spruce forest. These authors suggest that regional logging history may have created a forest landscape that is unable to support long-term occupancy of American Three-toed Woodpecker, and describe parallel losses of old growth forest bird species in Fennoscandia. They recommend that within these boreal systems, conservation strategies must extend beyond a network of protected areas to incorporate management strategies in unprotected regions that can benefit old-growth dependent species and help to prevent further declines.

Five papers expand our knowledge base for Rusty Blackbirds (*Euphagus carolinus*). Since Russ Greenberg and Sam Droege [10] identified this boreal wetland species as deserving enhanced study in 1999, much effort has contributed to better documentation of the bird's life history, habitat associations, and management concerns. Wilson et al.'s [11] assessment of genetic structure illuminated a continental population divide and a distinct group of Rusty Blackbirds in Newfoundland. The authors noted several landscape-scale genomic and demographic patterns, including an east–west partition consistent with separation during the last glacial maximum, a possible contact zone in Ontario, and limited gene flow between birds breeding in western and eastern North America. This evidence collectively suggests that current genomic variation in Rusty Blackbird populations results from both historical and contemporary processes, and that the capacity of Rusty Blackbirds to respond to rapid environmental change may be limited.

With Newfoundland as the home of one of only two known subspecies (*E. c. nigrans*), environmental changes and land management actions there may have important implications for the continent's eastern Rusty Blackbirds. Manson et al. [12] used forest resource inventory data and red squirrel presence to model Rusty Blackbird habitat occupancy in western Newfoundland. This 108,860 km^2 island off Canada's Atlantic coast hosts some of the highest known breeding densities of Rusty Blackbirds, and its landscape includes abundant old growth forest, which is rare in other parts of the breeding range. As documented elsewhere across boreal North America, lakes, ponds, rivers, streams, and bogs were strong predictors of Rusty Blackbird occupancy. Red squirrels are known nest predators, but did not appear to have a strong influence on blackbird habitat occupancy. The high breeding density, slower rate of decline, and genetic uniqueness of Newfoundland's Rusty Blackbird population justify further attention to the full life cycle ecology of these birds in comparison to populations on the North American mainland.

The use of wetlands for foraging is a hallmark of Rusty Blackbird ecology, and the birds are well-known both for capturing aerial insects and for foraging on vegetated edges of beaver ponds and similar wetland sites. Pachomski et al. [13] assessed aquatic invertebrate communities in occupied wetlands and found that prey abundance is positively associated with wetland occupancy by Rusty Blackbirds. The authors evaluated both remotely sensed and field-documented habitat variables and determined that the strongest predictors of Rusty Blackbird occupancy were wetlands with the lowest mud cover. Aquatic invertebrate abundance is another important predictor of wetland occupancy by Rusty Blackbirds, which is expected given the prevalence of those items in the bird's diet. In addition to landscape variables, fine-scale features can be useful in identifying potential foraging habitat and prioritizing areas for protection of Rusty Blackbird breeding habitat. Rusty Blackbird detection was highest during the chick-rearing stage, when adults were less secretive than during the earlier nest-building and incubation periods. The authors' survey methodology included a thirty-minute count period, which is three times longer than is typically used in songbird point counts. This longer sampling window may have implications for improving detection in understudied areas of the bird's range such as interior Canada, and for detection and occupancy modeling of other cryptic or hard to detect species.

When a young boreal bird leaves the nest, it risks potential injury or death as it first navigates the world. Thus, understanding survival, habitat use, and post-fledging ecology of both young-of-year and adult birds is important. Wohner et al. [14] provide novel information about Rusty Blackbird use of wetlands during the post-nesting period. Fledglings were likely to be closer to streams and in denser stands of sapling conifers and mixed-woods than adult birds. Further, while it is heartening that juvenile Rusty Blackbirds had higher survival than many other similar-sized species, survival of adults post-fledging was relatively low, suggesting the species may experience a gauntlet of predation or other stressors through which some adults do not pass. While the cause of this lower adult survival is unknown, this new information may help direct further research on other post-fledging boreal birds and their habitats.

An exciting development involving citizen science has implications for wintering habitat of boreal birds. Evans et al. [15] used public data from the Rusty Blackbird Winter Blitz and the eBird database to model flock size and patterns of space use in the south-eastern United States. They found that forest cover was the driver of Rusty Blackbird occupancy of floodplain forests, while average minimum temperature was also a factor in flock distribution. The research team found large flocks are best supported by a specific environmental niche, while smaller flocks could occupy a broader range of environmental conditions. Identification of several hotspots of wintering Rusty Blackbirds in the Lower Mississippi Alluvial Valley, the South Atlantic Coastal Plain, and the Black Belt Prairie can direct future conservation and habitat management efforts. Southern wintering grounds have experienced significant habitat loss, fragmentation and conversion to other land use types, and for a social species like the Rusty Blackbird, access to large wetland forests can help the birds maintain bonds as well as survive, if not thrive, to return to the boreal forest for another summer.

Conflicts of Interest: The authors declare no conflict of interest.

References

1. Flint, V.; Boehme, R.; Kostin, Y.; Kuznetsov, A. *A Field Guide to Birds of the USSR*; Princeton University Press: Princeton, NJ, USA, 1984.
2. Hagemeijer, W.; Blair, M. (Eds.) *The EBCC Atlas of European Breeding Birds*; T & A D Poyser: London, UK, 1997.
3. Sibley, D. *The Sibley Guide to Birds*; Alfred A. Knopf: New York, NY, USA, 2000.
4. Rosenberg, K.; Dokter, A.; Blancher, P.; Sauer, J.; Smith, A.; Smith, P.; Stanton, J.; Panjabi, A.; Helft, L.; Parr, M.; et al. Decline of the North American avifauna. *Science* **2019**, *366*, 120–124. [CrossRef] [PubMed]
5. Inger, R.; Gregory, R.; Duffy, J.; Stott, I.; Vorisek, P.; Gaston, K. Common European birds are declining rapidly while less abundant species' numbers are rising. *Ecol. Lett.* **2015**, *18*, 28–36. [CrossRef]
6. Ralston, J.; DeLuca, W. Conservation Lessons from the Study of North American Boreal Birds at Their Southern Periphery. *Diversity* **2020**, *12*, 257. [CrossRef]
7. Virkkala, R.; Aalto, J.; Heikkinen, R.; Rajasärkkä, A.; Kuusela, S.; Leikola, N.; Luoto, M. Can Topographic Variation in Climate Buffer against Climate Change-Induced Population Declines in Northern Forest Birds? *Diversity* **2020**, *12*, 56. [CrossRef]
8. Westwood, A.; Lambert, J.; Reitsma, L.; Stralberg, D. Prioritizing Areas for Land Conservation and Forest Management Planning for the Threatened Canada Warbler (*Cardellina canadensis*) in the Atlantic Northern Forest of Canada. *Diversity* **2020**, *12*, 61. [CrossRef]
9. Lamarre, V.; Tremblay, J. Occupancy of the American Three-Toed Woodpecker in a Heavily-Managed Boreal Forest of Eastern Canada. *Diversity* **2021**, *13*, 35. [CrossRef]
10. Greenberg, R.; Droege, S. On the decline of the Rusty Blackbird and the use of ornithological literature to document long-term population trends. *Conserv. Biol.* **1999**, *13*, 553–559. [CrossRef]
11. Wilson, R.; Matsuoka, S.; Powell, L.; Johnson, J.; Demarest, D.; Stralberg, D.; Sonsthagen, S. Implications of Historical and Contemporary Processes on Genetic Differentiation of a Declining Boreal Songbird: The Rusty Blackbird. *Diversity* **2021**, *13*, 103. [CrossRef]
12. Manson, K.; McDermott, J.; Powell, L.; Whitaker, D.; Warkentin, I. Assessment of Rusty Blackbird Habitat Occupancy in the Long Range Mountains of Newfoundland, Canada Using Forest Inventory Data. *Diversity* **2020**, *12*, 340. [CrossRef]
13. Pachomski, A.; McNulty, S.; Foss, C.; Cohen, J.; Farrell, S. Rusty Blackbird (*Euphagus carolinus*) Foraging Habitat and Prey Availability in New England: Implications for Conservation of a Declining Boreal Bird Species. *Diversity* **2021**, *13*, 99. [CrossRef]
14. Wohner, P.; Foss, C.; Cooper, R. Rusty Blackbird Habitat Selection and Survivorship during Nesting and Post-Fledging. *Diversity* **2020**, *12*, 221. [CrossRef]
15. Evans, B.; Powell, L.; Demarest, D.; Borchert, S.; Greenberg, R. Flock Size Predicts Niche Breadth and Focal Wintering Regions for a Rapidly Declining Boreal-Breeding Passerine, the Rusty Blackbird. *Diversity* **2021**, *13*, 62. [CrossRef]

Perspective

Conservation Lessons from the Study of North American Boreal Birds at Their Southern Periphery

Joel Ralston [1,*] **and William V. DeLuca** [2]

1   Department of Biology, Saint Mary's College, Notre Dame, IN 46556, USA
2   National Audubon Society, Amherst, MA 01075, USA; william.deluca@audubon.org
*   Correspondence: jralston@saintmarys.edu

Received: 22 May 2020; Accepted: 19 June 2020; Published: 24 June 2020

Abstract: Many North American boreal forest birds reach the southern periphery of their distribution in the montane spruce–fir forests of northeastern United States and the barren coastal forests of Maritime Canada. Because the southern periphery may be the first to be impacted by warming climates, these populations provide a unique opportunity to examine several factors that will influence the conservation of this threatened group under climate change. We discuss recent research on boreal birds in Northeastern US and in Maritime Canada related to genetic diversity, population trends in abundance, distributional shifts in response to climate change, community composition, and threats from shifting nest predators. We discuss how results from these studies may inform the conservation of boreal birds in a warming world as well as open questions that need addressing.

Keywords: range periphery; spruce–fir forests; climate change; range shift; community dynamics

1. Introduction

The North American boreal forest covers approximately 600 million ha of land across northern United States and Canada and is home to up to 3 billion breeding birds of more than 300 species [1]. Still a relatively undisturbed ecosystem across much of its extent, the boreal forest is perhaps the most productive ecosystem for birds in North America [2]. Yet the boreal forest faces considerable threats from various environmental pressures—perhaps the most important of which is climate change. High-latitude ecosystems such as boreal forests are expected to experience the greatest temperature changes in the coming decades [3], increased frequency of wildfires [4], and major changes in biodiversity [5]. As a result, boreal birds are among the most threatened bird communities by ongoing climate change [6,7]. This highlights the importance of understanding the ecology of boreal populations currently being impacted by climate change, and using that information to inform conservation action across the boreal.

The boreal forest reaches its southern and eastern peripheries in the low-lying wetlands and montane forests of northeastern United States and in coastal barren forests of maritime Canada [8,9] (Figure 1). These peripheral populations are likely to be the first to experience the ecological consequences of ongoing anthropogenic climate change. Southern periphery high-elevation populations are expected to be the first to be extirpated as habitats become unsuitable and species' distributions shift in response [10–12]. Additionally, as temperate species from further south shift into higher latitudes, southern boreal species will experience novel ecological interactions via competition, predation, and disease—each of which may impact population trends and community structure [12–15]. Lastly, southern boreal forests are impacted by recreation, land use change, development and disturbance associated with human populations at a much higher rate than the relatively undisturbed boreal forests further north [12,16–20]. For these reasons, we describe the southern boreal forest in northeastern United States and southern Canada as the current "battle ground" of environmental change for the boreal forest. Studies of ecological change and the impacts of climate change in these peripheral populations may therefore provide insight into how climate change will continue to impact boreal forests more broadly in the years to come. Here, we review approximately a decade of research in the southern periphery of the boreal forest, and synthesize several conservation lessons as well as ongoing questions related to the impacts of climate change on this community.

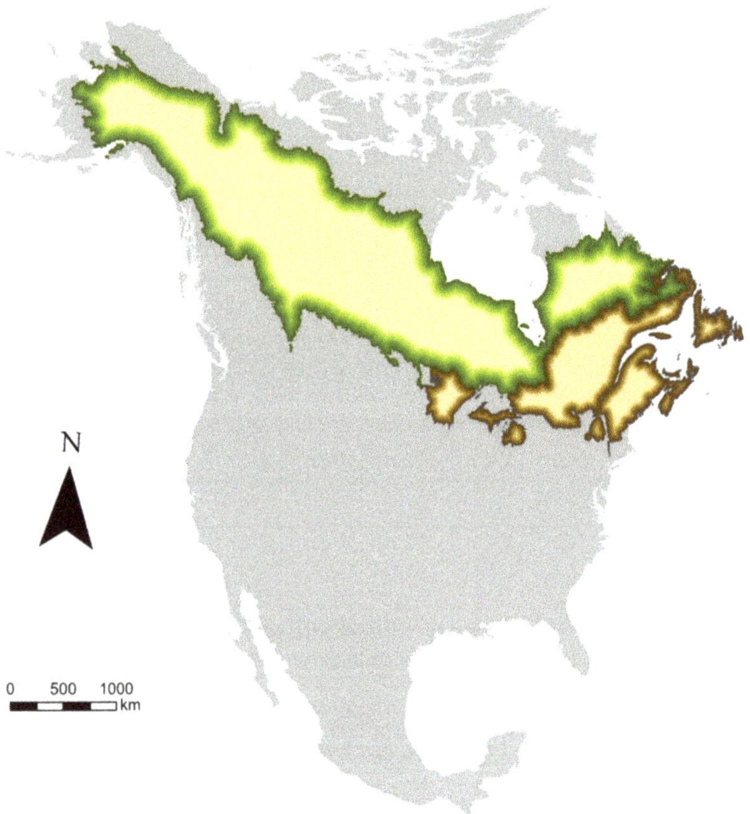

Figure 1. North American boreal forest (green) and a general approximation of its southern periphery (brown), where most of the studies reviewed here took place. The boreal forest and the southern periphery were identified by Bird Conservation Regions 4, 6, 7, 8, 12, and 14.

2. Peripheral Populations of Boreal Birds Are Genetically Unique, Threatened by Climate Change, and Declining

Because southern peripheral populations are geographically disjunct, they may represent an 'archipelago' of independently evolving populations isolated from gene flow and holding unique genetic diversity important to conserve under climate change [21,22]. In an analysis of microsatellite diversity, Ralston and Kirchman [11] found that southern peripheral populations of Blackpoll Warblers (*Setophaga striata*) in montane forests of northeastern United States may hold as much as 10% of the allelic richness in this species. An analysis of mitochondrial haplotypes across four species of boreal birds also found a periphery effect, with greater differences in genetic diversity between peripheral populations than between populations across the contiguous boreal forest [23]. This effect varied across species, with Blackpoll Warbler and Canada Jay (*Perisoreus canadensis*) showing a larger effect than Boreal Chickadee (*Poecile hudsonica*) and Yellow-bellied Flycatcher (*Empidonax flaviventris*), suggesting the isolating effect of peripheral populations may vary across species [23]. As populations decline at the southern periphery of the boreal, it is likely that this unique genetic diversity will be eroded [11,24].

In addition to southern peripheral populations, Newfoundland, an eastern peripheral isolate, has been shown to contain genetically unique populations of several boreal bird species including Canada Jay [25,26], Boreal Chickadee [27], Blackpoll Warbler [28], Gray-cheeked Thrush (*Catharus minimus*) [29], and Fox Sparrow (*Passerella iliaca*) [30]. Genetic divergence on Newfoundland may be due to isolation during the Pleistocene in a refugium located at the now submerged Grand Banks area east of modern day Newfoundland [25,27], or due to limited modern gene flow between Newfoundland and continental populations [23,28].

In addition to unique intraspecific genetic diversity observed in several species, the southern boreal forest is also home to an endemic bird species. The Bicknell's Thrush (*Catharus bicknelli*) breeds in upland spruce–fir forests in Northeastern United States, Quebec, New Brunswick, and Nova Scotia [31], a distribution closely matching the 'southern periphery' of other boreal forest species described above. While intraspecific divergence in other eastern boreal species has been dated to the Holocene, divergence between Bicknell's Thrush and its northern sister species, Gray-cheeked Thrush, has been dated to the Pleistocene [23,32]. Pleistocene divergence in these species likely occurred in separate glacial refugia, and may be reinforced by habitat niche differentiation or competition [29,33]. While deep genetic divergence in this *Catharus* species complex is unique among eastern boreal species studied thus far [23], Fitzerald et al. [29] suggest the possibility that similar cryptic species breaks may exist for other boreal bird species at the southern and eastern peripheries.

As the climate continues to warm, it is these unique peripheral populations that will be first impacted. Ralston and Kirchman [11] modeled predicted changes in climate suitability for 15 boreal forest birds under several climate change scenarios across the southern boreal. Suitability for all species was predicted to shift northward (mean 934.0 km under the most severe scenario), with 12 of 15 species predicted to be extirpated (99–100% decrease in suitable area) in New York, Vermont, and New Hampshire by 2080, resulting in a significant decline in boreal bird diversity in this region. While these simple correlative models used only climatic variables, further analysis with more sophisticated models has similarly projected declines in the suitability of boreal birds in northeastern United States in the coming decades. McGarigal et al. [34] used climate, habitat, and land use variables to calculate and project changes in Landscape Capability, an index reflecting the ability of the landscape to meet species' breeding season natural history requirements [35]. There was a projected 64% decline in Landscape Capability for Blackpoll Warbler across northeastern United States by 2080, along with a predicted upslope shift in distribution greater than 100 m.

Impacts from climate change on bird distributions and abundances are already being observed in this region. In a resurvey of birds along an altitudinal gradient in New York state [36], Kirchman and Van Keuren [37] found the abundance weighted mean altitude shifted upslope an average 82.8 m (*n* = 42 species) between 1974 and 2015. This result mirrors similar findings for long term upslope shifts in montane forest ecotones [38]. As climate change pushes species up in elevation, we expect populations to decline due to less area available at higher elevations [39], especially in the southern parts of their range where birds are already found near the tops of mountains. However, there appears to be variation in elevational shifts across sites and species. DeLuca and King [12] found that 9 out of 11 montane boreal species in the White Mountains shifted their lower elevational range boundary downslope on average 19 m between 1993 and 2009. This shift may have been a response to regrowth of spruce and fir following historic logging [12,40].

As climate change continues, boreal birds are declining. Continental-scale point count surveys typically used to assess long term population trends in birds are usually unsuitable for boreal species given the inaccessibility of their habitats [41–43], but the estimates that are available often show steep declines. Blackpoll Warblers, an emblematic bird of the boreal forest and historically perhaps the most abundant, has seen a 92% decline since the 1970s [44]. To estimate trends at the local and regional scales, numerous surveys exist in montane spruce–fir forests, or lowland boreal spruce bogs across the southern boreal, operated by conservation organizations, national forests, parks, and wildlife refuges [45–51]. Several of these programs have now been operated for over two decades, and most are showing a general pattern of declining bird abundances across the southern boreal. Vermont Center for Ecostudies' Mountain Birdwatch has documented a significant 10 year decline in 7 of the 10 species that it monitors in montane forests in northeastern United States and Canada [51]. The only species monitored that significantly increased in the Mountain Birdwatch dataset was Black-capped Chickadee (*Poecile atricapilla*), indicating the movement of low-elevation species into high-elevation boreal habitats and potentially a change in community structure [51]. Similarly, Glennon et al. [50] found that boreal birds generally declined in occupancy at lowland boreal wetland sites in New York over a decade of observation, while non-boreal species were more likely to increase in occupancy. Further, non-boreal species were more likely to colonize sites that boreal species disappeared from, again indicating changes in community composition [50]. Ralston et al. [42] combined 16 of these local and regional survey datasets to estimate population trends for 14 boreal bird species across the northeastern and upper Midwestern United States. They found that four species commonly considered indicator species of boreal spruce–fir forests, Yellow-bellied Flycatcher, Bicknell's Thrush, Blackpoll Warbler, and Magnolia Warbler (*Setophaga magnolia*), each showed significant declines across the study region. Further, species identified as spruce–fir specialists were more likely to be declining than those which also used other habitat types, indicating that pressures driving population declines in abundance in birds may be particularly strong in boreal spruce–fir forests.

In addition to showing general declines in spruce–fir birds, Ralston et al. [42] demonstrated considerable regional variation in trends within species. In five of the nine species with sufficient data for analysis in northeastern and upper Midwestern sites, trends were significantly different between regions. In three species the direction of change differed. Canada Jay and Swainson's Thrush (*Catharus ustulatus*) each declined in the upper Midwest while increasing in abundance in the Northeast; Red-breasted Nuthatch showed the opposite pattern, increasing in the upper Midwest while declining in the Northeast [42]. Similarly, all species showed variation in trend estimates across surveys within each region. At a local scale, Glennon et al. [50] also showed variation among sites in boreal bird community dynamics. Boreal birds were more likely to persist at sites with a greater amount of open Northern Peatland compared to sites with Boreal Upland Forest (habitat classifications following [9]). However, it is unlikely that these results apply to boreal species more typical of Boreal Upland Forests such as Blackpoll Warbler and Bicknell's Thrush [50]. Identifying the habitat factors associated with stable or increasing trends in boreal birds at the southern periphery could be very important for conservation practitioners. Cooperation among stakeholders to extend previous analyses to a greater number of sites, boreal habitat types, and regions throughout the southern boreal could be tremendously helpful for the conservation of boreal birds more broadly. The importance of support for targeted boreal monitoring programs cannot be overstated. Current population estimates for boreal species based on road-based surveys have proven to be inaccurate and often underestimate boreal population estimates [43].

3. Species Differ in Their Responses to Changing Environments, Creating a Challenge for Community Level Conservation

Drivers of species distributions often vary across species. For boreal birds, the extent to which habitat and/or climate limit distributions has major implications for the conservation and management of those species. For example, if a species' distribution is primarily limited by climate, focusing on climate refugia preservation might be the preferred approach; if a species was primarily limited by habitat, management of habitat characteristics might be the optimal approach. However, a challenge of species-focused approach is that management plans designed to conserve a single species may have limited benefits for other threatened species in the community. Several recent studies in habitat associations of boreal birds have shown that species differ in their habitat requirements and in the environmental factors that drive their occupancy [12,50,52–54]. Climatic factors were shown to be generally more important than landscape factors (i.e., connectivity, human footprint) to occupancy dynamics in low-elevation boreal birds, though species differed in what climate factors were most important [54]. Additionally, indirect effects of climate as mediated by species interactions (i.e., insect prey abundance) and vegetation structure may be important to dynamics of these species on relatively short and long time scales, respectively [54]. Similarly, a large proportion of the impact of climate on the distribution of montane boreal birds along an elevation gradient was indirect, mediated by vegetation, and again there was great variation in the strength and direction of direct in indirect effects of climate across species [52]. Together, these results demonstrate variation in habitat associations across boreal birds. These differences lead to regional variation in the structure of boreal bird communities along the southern periphery of the boreal, and ongoing climate change may drive changes in community composition [50,53].

While it has been previously demonstrated that species associations may be weak in southern boreal forests, creating challenges to community level conservation [55], recent studies on boreal birds suggest that ongoing climate change may exacerbate these challenges by driving changes in community composition [50,52–54]. For example, designing forest management practices to benefit a species of conservation concern, Bicknell's Thrush [56,57], may have relatively smaller impact on other declining species such as Blackpoll Warbler or Canada Jay [42] that appear to be more strongly and directly influenced by climate variables [52,53]. While protecting climate refugia where boreal communities can persist may be an important conservation strategy [6,50], conserving the boreal community in its current form will become increasingly challenging. Species of conservation concern may be shuffled into non-analogous communities not currently present in the landscape, requiring unique conservation strategies [58]. Conservation practitioners focused on boreal species will therefore need to clearly define whether the objectives of management efforts are to protect individual species or larger assemblages of boreal birds.

4. As Communities Reshuffle, Species Face New Biotic Interactions—Some of Which May Have Unexpected Outcomes

As climate change drives changes in community composition, species will be faced with novel ecological interactions and challenges. For boreal birds at their southern periphery, increased contact with congeneric competitors may be increasingly important with climate change. Southern limits of northern species and, by extension, low-elevational limits of high-elevation species, are often determined by biotic interactions, mainly competition [59–61]. Many of the boreal species discussed thus far have more southerly or low-elevation relatives with which they compete [61,62], and that may be shifting into historically boreal sites [12,50,51]. In recent decades, as Bicknell's Thrush has been declining, the congeneric Swainson's Thrush has been shifting upslope and increasing in abundance, followed by Hermit Thrush (*Catharus guttata*), a historically low-elevation competitor species which can now be observed at many high-elevation sites [12,37,51,62]. Similarly, Boreal Chickadee may be being outcompeted and replaced by Black-capped Chickadees in many places [51]. In the White Mountains, warbler species typically associated with lower-elevation mixed forests shifted on average upslope into boreal habitats while high-elevation boreal warblers such as Blackpoll Warbler, Yellow-rumped Warbler (*Setophaga coronata*), and Nashville Warbler (*Oreothlypis ruficapilla*) have shifted downslope [12], increasing the likelihood of competition among these species. While warblers are a classic example of foraging niche differentiation to reduce competition [63], it is unclear how these high-elevation boreal species will respond to novel competitors.

Another possibility, as climate change pushes closely related species into contact, is hybridization. Secondary contact of related species following environmental change can result in widespread introgression, loss of genetic diversity from a parent species, and even speciation reversal or the collapse of two lineages into one [64–66]. This can create additional conservation concern, especially for rare species that come into contact and hybridize with a more abundant species [67]. However, hybridization can be a tricky conservation topic, as it may not always be clear whether hybrids are naturally occurring or the result of anthropogenic disturbance [68]. As climate change reshuffles species distributions and community composition, it is possible that there might be increased opportunities for hybridization, especially at range peripheries, where species' distributions abut one another. FitzGerald et al. [32] recently documented hybridization between Bicknell's Thrush and its northern sister species Gray-cheeked Thrush. The hybrid was phenotypically Gray-cheeked Thrush and was captured in southern Labrador within the Gray-cheeked Thrush's distribution, but held a Bicknell's Thrush mitochondrial haplotype [32]. Further analysis of both species has revealed low levels of genetic admixture within Bicknell's Thrush populations [29]. While hybridization between these species has been suggested in the past [69], this is the first confirmed case. While this likely represents a natural case of hybridization, it is unclear whether anthropogenic climate change may increase the frequency of hybridization over time, either through changes in distribution or migratory behavior (i.e., late-migrating Gray-cheeked Thrush, breeding with Bicknell's Thrush, as they migrate through Bicknell's Thrush territories further south) [32].

An important biotic interaction currently impacting peripheral boreal populations is the climate-driven introduction and spread of Red Squirrels (*Tamiasciurus hudsonicus*) an avian nest predator. Red Squirrels may be shifting downslope, opposite expectations given climate change [70] and are impacting nesting success of boreal birds in montane habitats [71]. Nest survival in Blackpoll Warblers and Bicknell's Thrush is highest at elevations with the lowest Red Squirrel abundance, and in years with lower squirrel abundance [71]. Similarly, the introduction and spread of Red Squirrels on Newfoundland has been coincident with the steep decline in Newfoundland Gray-cheeked Thrush populations [15,72], and Gray-cheeked Thrush were more likely to be detected on point counts where Red Squirrels were absent [32]. An interesting aspect of this is that Red Squirrels appear to be co-distributed with boreal birds elsewhere across the contiguous boreal forest with seemingly lower impacts on population abundance than at the southern and eastern peripheries of the boreal. This suggests that biotic interactions introduced through climate change can have unexpected consequences. Further work needs to be performed to understand how differing ecological contexts in peripheral populations and throughout the boreal influence the outcomes of these interactions.

5. Conclusions

Boreal forests are one of the most threatened ecosystems under climate change. At the southern periphery of the boreal forest, where climate change is already having an impact, boreal birds are declining in abundance and genetic diversity [24,42,51], and boreal community composition is changing [50,53]. We can take several conservation lessons from the studies of these peripheral boreal bird communities. First, species-specific responses to climate change may require individualized management plans, or flexible conservation that accounts for and is flexible to non-analogous communities. Second, as communities reshuffle, new biotic interactions are likely, and their outcomes may be unique to this new ecological context, providing additional challenges for conservation. Continued and expanded monitoring of boreal forest bird communities will be essential to detect further declines and understanding consequences of novel species assemblages. Future coordination of research and monitoring efforts across boreal regions and habitat types, and among conservation stakeholders, will help us to understand why trends in boreal birds vary regionally and what factors promote local increases in abundance and will be crucial for the conservation of boreal birds.

Author Contributions: Conceptualization, J.R. and W.V.D.; writing—original draft preparation, J.R.; writing—review and editing, W.V.D. All authors have read and agreed to the published version of the manuscript.

Funding: This research received no external funding.

Acknowledgments: This manuscript was developed from a talk given by the authors in the "Conservation and Management of Boreal Birds in a Changing Climate" symposium organized by Drs. D. Stralberg, S. Matsuoka, and J. Tremblay at the 2019 meeting of American Ornithological Society. We thank the organizers and the participants in that symposium for feedback and discussion on the topic. Additionally, we thank our many collaborators over the previous decade who have taught us a great deal about boreal birds.

Diversity **2020**, *12*, 257

Conflicts of Interest: The authors declare no conflict of interest.

References

1. Wells, J.; Stralberg, D.; Childs, D. *Boreal Forest Refuge: Conserving North America's Bird Nursery in the Face of Climate Change*; Boreal Songbird Initiative: Seattle, WA, USA, 2018.
2. Blancher, P.; Wells, J. *The Boreal Forest Region: North. America's Bird Nursery*; Canadian Boreal Initiative and Boreal Songbird Initiative: Ottawa, ON, Canada, 2005; pp. 1–12.
3. IPCC. *Climate Change 2013: The Physical Science Basis. Contribution of Working Group I to the Fifth Assessment Report of the Intergovernmental Panel on Climate Change*; Stocker, T.F., Qin, D., Plattner, G.-K., Tignor, M., Allen, S.K., Boschung, J., Nauels, A., Xia, Y., Bex, V., Midgley, P.M., Eds.; Cambridge University Press: Cambridge, UK; New York, NY, USA, 2018; 1535p.
4. Walker, X.J.; Baltzer, J.L.; Cumming, S.G.; Day, N.J.; Ebert, C.; Goetz, S.; Johnstone, J.F.; Potter, S.; Rogers, B.M.; Schuur, E.A.; et al. Increasing wildfires threaten historic carbon sink of boreal forest soils. *Nature* **2019**, *572*, 520–525. [CrossRef]
5. Nelson, T.A.; Coops, N.C.; Wulder, M.A.; Perez, L.; Fitterer, J.; Powers, R.; Fontana, F. Predicting climate change impacts to the Canadian boreal forest. *Diversity* **2014**, *6*, 133–157. [CrossRef]
6. Stralberg, D.; Berteaux, D.; Drever, C.R.; Drever, M.; Naujokaitis-Lewis, I.; Schmiegelow, F.K.A.; Tremblay, J.A. Conservation planning for boreal birds in a changing climate: A framework for action. *Avian Conserv. Ecol.* **2019**, *14*, 13. [CrossRef]
7. Bateman, B.L.; Wilsey, C.; Taylor, L.; Wu, J.; LeBaron, G.S.; Langham, G.M. North American birds require mitigation and adaptation to reduce vulnerability to climate change. *Conserv. Sci. Pract.* **2019**, in press.
8. Damman, A.W.H. An ecological subdivision of the island of Newfoundland. In *Biogeography and Ecology of the Island of Newfoundland*; South, G.R., Ed.; Dr. W. Junk Publishers: The Hague, The Netherlands, 1983; pp. 163–206.
9. Anderson, M.G.; Clark, M.; Ferree, C.E.; Jospe, A.; Olivero Sheldon, A.; Weaver, K.J. *Northeast. Habitat Guides: A Companion to the Terrestrial and Aquatic Habitat Maps*; The Nature Conservancy, Eastern Conservation Science: Boston, MA, USA, 2013.
10. Zukerburg, B.; Woods, A.M.; Porter, W.F. Poleward shufts in breeding bird distributions in New York State. *Glob. Chang. Biol.* **2009**, *15*, 1866–1883. [CrossRef]
11. Ralston, J.; Kirchman, J.J. Predicted range shifts in North American boreal forest birds and the effect of climate change on genetic diversity in blackpoll warblers (*Setophaga striata*). *Conserv. Genet.* **2013**, *14*, 543–555. [CrossRef]
12. DeLuca, W.V.; King, D.I. Montane birds shift downslope despite recent warming in the northern Appalachian Mountains. *J. Ornithol.* **2017**, *158*, 493–505. [CrossRef]
13. Crowl, T.A.; Crist, T.O.; Parmenter, R.R.; Belovsky, G.; Lugo, A.E. The spread of invasive species and infectious disease as drivers of ecosystem change. *Front. Ecol. Environ.* **2008**, *6*, 238–246. [CrossRef]
14. Fuller, T.; Bensch, S.; Müller, I.; Novembre, J.; Pérez-Tris, J.; Ricklefs, R.E.; Smith, T.B.; Waldenström, J. The ecology of emerging infectious diseases in migratory birds: An assessment of the role of climate change and priorities for future research. *EcoHealth* **2012**, *9*, 80–88. [CrossRef]
15. Whitaker, D.M. The colonization of Newfoundland by red squirrels (*Tamasciurus hudsonicus*). *Osprey Nat. J. Nfld. Labrador* **2015**, *46*, 23–29.
16. Imbeau, L.; Mönkkönen, M.; Desrochers, A. Long-term effects of forestry on birds of the eastern Canadian boreal forests: A comparison with Fennoscandia. *Conserv. Biol.* **2001**, *15*, 1151–1162. [CrossRef]
17. Rimmer, C.C.; McFarland, K.P.; Evers, D.C.; Miller, E.K.; Aubry, Y.; Busby, D.; Taylor, R.J. Mercury concentrations in Bicknell's Thrush and other insectivorous passerines in montane forests of Northeastern North America. *Ecotoxicology* **2005**, *14*, 223–240. [CrossRef] [PubMed]
18. Venier, L.A.; Thompson, I.D.; Fleming, R.; Malcolm, J.; Aubin, I.; Trofymow, J.A.; Langor, D.; Sturrock, R.; Patry, C.; Outerbridge, R.O.; et al. Effects of natural resource development on the terrestrial biodiversity of Canadian boreal forest. *Environ. Rev.* **2014**, *22*, 457–490. [CrossRef]
19. Zlonis, E.J.; Niemi, G.J. Avian communities of managed and wilderness hemiboreal forests. *For. Ecol. Manag.* **2014**, *328*, 26–34. [CrossRef]

20. DeLuca, W.V.; King, D.I. Influence of hiking trails on montane birds. *J. Wildlife Manag.* **2014**, *78*, 494–502. [CrossRef]

21. Hampe, A.; Petit, R.J. Conserving biodiversity under climate change: The rear edge matters. *Ecol. Lett.* **2005**, *8*, 461–467. [CrossRef] [PubMed]

22. Kirchman, J.J.; Ralston, J. The Adirondack archipelago. *Adirondack J. Environ. Stud.* **2015**, *20*, 5.

23. Ralston, J.; FitzGerald, A.M.; Burg, T.M.; Starkloff, N.C.; Whitaker, D.M.; Warkentin, I.G.; Norris, D.R.; Kirchman, J.J. Comparative phylogeographic analysis suggests a shared history among eastern North American boreal forest birds. *Glob. Ecol. Biogeog.* Under review.

24. Kirchman, J.J.; Ross, A.M.; Johnson, G. Historical decline of genetic diversity in a range-periphery population of Spruce Grouse (*Falcipennis canadensis*) inhabiting the Adirondack Mountains. *Conserv. Genet.* **2020**, *21*, 373–380. [CrossRef]

25. Van Els, P.; Cicero, C.; Klicka, J. High latitude and high genetic diversity: Phylogeography of a widespread boreal bird, the gray jay (*Perisoreus canadensis*). *Mol. Phylogenet. Evol.* **2012**, *63*, 456–465. [CrossRef]

26. Dohms, K.M.; Graham, B.A.; Burg, T.M. Multilocus genetic analyses and spatial modeling reveal complex population structure and history in a widespread resident North American passerine (*Perisoreus canadensis*). *Ecol. Evol.* **2017**, *7*, 9869–9889. [CrossRef] [PubMed]

27. Lait, L.A.; Burg, T.M. When east meets west: Population structure of a high-latitude resident species, the boreal chickadee (*Poecile hudsonicus*). *Heredity* **2013**, *111*, 321–329. [CrossRef] [PubMed]

28. Ralston, J.; Kirchman, J.J. Continent-scale genetic structure in a boreal forest migrant, the blackpoll warbler (*Steophaga striata*). *Auk* **2012**, *129*, 467–478.

29. FitzGerald, A.M.; Weir, J.; Ralston, J.; Warkentin, I.G.; Whitaker, D.M.; Kirchman, J.J. Genetic structure and biogeographical history of the Bicknell's Thrush/Gray-cheeked Thrush species complex. *Auk* **2020**, *137*, 1–20. [CrossRef]

30. Zink, R.M. The geography of mitochondrial DNA variation, population structure, hybridization, and species limits in the fox sparrow (*Passerella iliaca*). *Evolution* **1994**, *48*, 96–111.

31. Townsend, J.M.; McFarland, K.P.; Rimmer, C.C.; Ellison, W.G.; Goetz, J.E. Bicknell's Thrush (*Catharus bicknelli*), version 1.0. In *Birds of the World*; Rodewald, P.G., Ed.; Cornell Lab of Ornithology: Ithaca, NY, USA, 2020. [CrossRef]

32. FitzGerald, A.M.; Whitaker, D.M.; Ralston, J.; Kirchman, J.J.; Warkentin, I.G. Taxonomy and distribution of the imperiled Newfoundland Gray-cheeked Thrush, *Catharus minimus minimus. Avian Conserv. Ecol.* **2017**, *12*, 10. [CrossRef]

33. FitzGerald, A.M. Division within the North American boreal forest: Ecological niche divergence between the Bicknell's Thrush (*Cathatus bicknelli*) and Gray-cheeked Thrush (*C. minimus*). *Ecol. Evol.* **2017**, *7*, 5285–5295. [CrossRef] [PubMed]

34. Designing Sustainable Landscapes: Modeling Focal Species. Report to the North Atlantic Conservation Cooperative, US Fish and Wildlife Service, Northeast Region. Available online: http://jamba.provost.ads.umass.edu/web/lcc/dsl/technical/DSL_documentation_species.pdf (accessed on 22 May 2020).

35. Loman, Z.G.; DeLuca, W.V.; Harrison, D.; Loftin, C.S.; Wood, P.B. Landscape capability models as a tool to predict fine-scale forest bird occupancy and abundance. *Landsc. Ecol.* **2018**, *33*, 77–91. [CrossRef]

36. Able, K.P.; Noon, B.R. Avian community structure along elevational gradients in the Northeastern United States. *Oecologia* **1979**, *26*, 275–294. [CrossRef]

37. Kirchman, J.J.; Van Keuren, A.E. Altitudinal range shifts of birds at the southern periphery of the boreal forest: 40 years of change in the Adirondacks Mountains. *Wilson J. Ornithol.* **2017**, *129*, 742–753. [CrossRef]

38. Beckage, B.; Osborne, B.; Gavin, D.G.; Pucko, C.; Siccama, T.; Perkins, T. A rapid upward shift of the forest ecotone during 40 years of warming in the Green Mountains of Vermont. *Proc. Natl. Acad. Sci. USA* **2008**, *105*, 4197–4202. [CrossRef] [PubMed]

39. Elsen, P.R.; Tingley, M.W. Global mountain topography and the fate of montane species under climate change. *Nat. Clim. Chang.* **2015**, *5*, 772–776. [CrossRef]

40. Foster, J.R.; D'Amato, A.W. Montane forest ecotones moved downslope in northeastern USA in spite of warming between 1984 and 2011. *Glob. Chang. Biol.* **2015**, *21*, 4497–4507. [CrossRef] [PubMed]

41. Hanowski, J.A.; Niemi, G.J. A comparison of on- and off-road bird counts: Do you need to go off road to count birds accurately? *J. Field Ornithol.* **1995**, *66*, 469–483.

42. Ralston, J.; King, D.I.; DeLuca, W.V.; Niemi, G.J.; Glennon, M.J.; Scarl, J.C.; Lambert, J.D. Analysis of combined data sets yields trend estimates for vulnerable spruce-fir birds in northern United States. *Biol. Conserv.* **2015**, *187*, 270–278. [CrossRef]

43. Sólymos, P.; Toms, J.D.; Matsuoka, S.M.; Cumming, S.G.; Barker, N.K.; Thogmartin, W.E.; Stralberg, D.; Crosby, A.D.; Dénes, F.V.; Haché, S.; et al. Lessons learned from comparing spatially explicit models and the Partners in Flight approach to estimate population sizes of boreal birds in Alberta, Canada. *Condor* **2020**. [CrossRef]

44. Partners in Fight Science Committee: Partners in Flight Landbird Conservation Plan: 2016 Revision for Canada and Continental United States. Available online: https://www.partnersinflight.org/wp-content/uploads/2016/08/pif-continental-plan-final-spread-single.pdf. (accessed on 22 May 2020).

45. Howe, R.H.; Roberts, L.J. *Sixteen Years of Habitat-Based Bird Monitoring in the Nicolet National Forest*; USDA Forest Service General Technical Report PSW-GTR-191; USDA Forest Service: Washington, DC, USA, 2005.

46. US Forest Service. *White Mountain National Forest, Monitoing and Valuation Report*; US Department of Argiculture: Washinginton, DC, USA, 2006.

47. Johnson, B. *The Ottawa National Forest Breeding Bird Census, an Analysis of Twenty Years of Data, 1992–2011*; Report Submitted to Ottawa National Forest; USDA Forest Service: Ironwood, MI, USA, 2012.

48. Zlonis, E.J.; Grinde, A.; Bednar, J.; Niemi, G.J. *Summary of Breeding Bird Trends in the Chippewa and Superior National Forests of Minnesota- 1995–2013*; NRRI Technical Report NRRI/TR-2013/36; University of Minnnesota: Duluth, MN, USA, 2013.

49. Faccio, S.D.; Mitchell, B.R. *Breeding Landbird Monitoring in the Northeast*; Temperate Network, 2013 Summary Report; Natural Resource Data Series NPS/NETN/NRDS-2014/630; National Park Service: Fort Collins, CO, USA, 2014.

50. Glennon, M.J.; Langdon, S.F.; Rubenstein, M.A.; Cross, M.S. Temporal changes in avian community composition in lowland conifer habitats at the southern edge of the boreal zone in the Adirondack Park, NY. *PLoS ONE* **2019**, *14*, e0220927. [CrossRef]

51. Hill, J.M. *The State of the Mountain Birds Report: 2020*; Vermont Center for Ecostudies: White River Junction, VT, USA, 2020.

52. Duclos, T.R.; DeLuca, W.V.; King, D.I. Direct and indirect effects of climate on bird abundance along elevational gradients in the Northern Appalachian Mountains. *Divers. Distrib.* **2019**, *25*, 1670–1683. [CrossRef]

53. Ralston, J.; FitzGerald, A.M.; Scanga, S.E.; Kirchman, J.J. Observations of habitat associations in boreal forest birds and the geographic variation in bird community composition. *Wilson J. Ornithol.* **2019**, *131*, 12–23. [CrossRef]

54. Glennon, M.J.; Langdon, S.F.; Rubenstein, M.A.; Cross, M.S. Relative contribution of climate and non-climate drivers in determining dynamic rates of boreal birds at the edge of their range. *PLoS ONE* **2019**, *14*, e0224308. [CrossRef]

55. Niemi, G.J.; Hanowski, J.M.; Lima, A.R.; Nicholls, T.; Weiland, N. A critical analysis on the use of indicator species in management. *J. Wildlife Manag.* **1997**, *61*, 1240–1252. [CrossRef]

56. Aubry, Y.; Desrochers, A.; Seutin, G. Regional patterns of habitat use by a threatened forest bird, the Bicknell's Thrush (*Catharus bicknelli*), in Quebec. *Can. J. Zool.* **2016**, *94*, 301–309. [CrossRef]

57. Chisholm, S.E.; Leonard, M.L. Effect of forest management on a rare habitat specialist, the Bicknell's Thrush (*Catharus bicknelli*). *Can. J. Zool.* **2008**, *86*, 217–223. [CrossRef]

58. Keith, S.A.; Newton, A.C.; Herbert, R.J.H.; Morecroft, M.D.; Bealey, C.E. Non-analogous community formation in response to climate change. *J. Nat. Conserv.* **2009**, *17*, 228–235. [CrossRef]

59. MacArthur, R.H. *Geographical Ecology: Patterns in the Distribution of Species*; Harper and Row: New York, NY, USA, 1972.

60. Brown, J.H.; Lomolino, M.V. *Biogeography*, 2nd ed.; Sinauer Associates Inc.: Sunderland, MA, USA, 1998.

61. Freeman, B.G.; Montgomery, G. Interspecific aggression by the Swainson's Thrush (*Catharus ustulatus*) may limit the distribution of the threatened Bicknell's Thrush (*Catharus bicknelli*) in the Adirondack Mountains. *Condor* **2016**, *118*, 169–178. [CrossRef]

62. Noon, B.R. The distribution of an avian guild along a temperate elevational gradient: The importance and expression of competition. *Ecol. Monogr.* **1981**, *51*, 105–124. [CrossRef]

63. MacArthur, R.H. Population ecology of some warblers of northeastern coniferous forests. *Ecology* **1958**, *39*, 599–619. [CrossRef]

64. Gill, F.B. Local cytonuclear extinction of the golden-winged warbler. *Evolution* **1997**, *51*, 519–525. [CrossRef]
65. Vallender, R.; Van Wilgenburg, S.L.; Bulluck, L.P.; Roth, A.; Canterbury, R.; Larkin, J.; Fowlds, R.M.; Lovette, I.J. Extensive rangewide mitochondrial introgression indicated substantial cryptic hybridization in the golden-winged warbler (*Vermivora chrysoptera*). *Avian Conserv. Ecol.* **2009**, *4*, 4. [CrossRef]
66. Kearns, A.M.; Restani, M.; Szabo, I.; Schrøder-Nielsen, A.; Kim, J.A.; Richardson, H.M.; Marzluff, J.M.; Fleischer, R.C.; Johnsen, A.; Omland, K.E. Genomic evidence of speciation reversal in ravens. *Nat. Commun.* **2018**, *9*, 1–13. [CrossRef]
67. Rhymer, J.M.; Simberloff, D. Extinction by hybridization and introgression. *Annu. Rev. Ecol. Syst.* **1996**, *27*, 83–109. [CrossRef]
68. Allendorf, F.W.; Leary, R.F.; Spurell, P.; Wenburg, J.K. The problems with hybrids: Setting conservation guidelines. *Trends Ecol. Evol.* **2001**, *16*, 613–622. [CrossRef]
69. Marshall, J.T. *The Gray-Cheeked Thrush, Catharus minimus, and Its New England Subspecies, Bicknell's thrush, Catharus minimus bicknelli. No. 28*; Nuttall Ornithological Club: Cambridge, MA, USA, 2001.
70. Morelli, T.L.; Hallworth, M.T.; Duclos, T.; Siren, A.; Ells, A.; Faccio, S.; McFarland, K.; Nislow, K.; Ralston, J.; Ratnaswamy, M.; et al. Habitat causes lags in response to climate change for a climate-vulnerable species. *Glob. Ecol. Biogeog.* Under review.
71. Hallworth, M.; Siren, A.; DeLuca, W.V.; Duclos, T.; McFarland, K.P.; Hill, J.; Rimmer, C.C.; Morelli, T.L. Boom and bust: Pulsed resources drive distribution dynamics in forested food webs. *Glob. Ecol. Biogeog.* Under review.
72. Whitaker, D.M.; Taylor, P.D.; Warkentin, I.G. Gray-cheeked Thrush (*Catharus minimus minimus*) distribution and habitat use in a montane forest landscape of western Newfoundland, Canada. *Avian Conserv. Ecol.* **2015**, *10*, 4. [CrossRef]

Article

Can Topographic Variation in Climate Buffer against Climate Change-Induced Population Declines in Northern Forest Birds?

Raimo Virkkala [1,*], Juha Aalto [2,3], Risto K. Heikkinen [1], Ari Rajasärkkä [4], Saija Kuusela [1], Niko Leikola [1] and Miska Luoto [3]

[1] Finnish Environment Institute, Biodiversity Centre, Latokartanonkaari 11, FI-00790 Helsinki, Finland; risto.heikkinen@ymparisto.fi (R.K.H.); saija.kuusela@ymparisto.fi (S.K.); niko.leikola@ymparisto.fi (N.L.)
[2] Finnish Meteorological Institute, FI-00101 Helsinki, Finland; juha.aalto@fmi.fi
[3] Department of Geosciences and Geography, University of Helsinki, P.O. Box 64, FI-00014 Helsinki, Finland; miska.luoto@helsinki.fi
[4] Metsähallitus, National Parks Finland, P.O. Box 81, FI-90101 Oulu, Finland; ari.rajasarkka@metsa.fi
* Correspondence: raimo.virkkala@ymparisto.fi

Received: 19 December 2019; Accepted: 31 January 2020; Published: 1 February 2020

Abstract: Increased attention is being paid to the ecological drivers and conservation measures which could mitigate climate change-induced pressures for species survival, potentially helping populations to remain in their present-day locations longer. One important buffering mechanism against climate change may be provided by the heterogeneity in topography and consequent local climate conditions. However, the buffering capacity of this topoclimate has so far been insufficiently studied based on empirical survey data across multiple sites and species. Here, we studied whether the fine-grained air temperature variation of protected areas (PAs) affects the population changes of declining northern forest bird species. Importantly to our study, in PAs harmful land use, such as logging, is not allowed, enabling the detection of the effects of temperature buffering, even at relatively moderate levels of topographic variation. Our survey data from 129 PAs located in the boreal zone in Finland show that the density of northern forest species was higher in topographically heterogeneous PAs than in topographically more homogeneous PAs. Moreover, local temperature variation had a significant effect on the density change of northern forest birds from 1981–1999 to 2000–2017, indicating that change in bird density was generally smaller in PAs with higher topographic variation. Thus, we found a clear buffering effect stemming from the local temperature variation of PAs in the population trends of northern forest birds.

Keywords: boreal; buffering; climate change; forest bird; macroclimate; population decline; protected areas; topographic heterogeneity

---

1. Introduction

Climate change is a major threat to biodiversity [1,2], increasingly affecting present-day species populations and communities [3–6]. Upslope shifts in the distribution of species have already been observed with rising temperatures both in temperate [7,8] and tropical regions [9,10], although the elevational shifts may vary considerably in terms of direction and magnitude among species [8]. Changes in climate are projected to cause further poleward and upward species range shifts in the future [2,11]; but, in many areas rapidly changing conditions will challenge the ability of species to track the spatial reorganization of suitable locations [12–14]. Thus, climate-smart conservation strategies are required to identify landscape characteristics which best support the range-shifting and persistence of species populations [15,16]. One promising strategy is the prioritization of areas with notable fine-grain heterogeneity in local climates caused by variation in topography (i.e., topoclimatic heterogeneity) [16–19]. The conservation of landscapes with high topoclimatic variation allows species to move shorter distances while tracking suitable conditions. Thereby, it may facilitate the extended persistence of trailing-edge (the contracting or retreating edge of range) populations by providing refuges and holdouts with favorable local climate conditions [20–23]. Thus, variation in topoclimates provides a potentially important buffering mechanism against the impacts of climate change on biodiversity [17].

However, although the importance of topoclimatic variation for biodiversity preservation is intuitively credible, support for it largely comes from theoretical and model-based studies, or climatological assessments of variability in local climatic conditions [21,24–26]. In essence, the evidence on the benefits of topoclimatic variation emerging from species surveys conducted across multiple sites or protected areas (PAs) is still scarce (see, however, [19]). Monitoring studies of species trends in topographically diverse landscapes both in PAs and outside them have revealed contrasting trends, including upslope as well as downslope expansion of populations, increases in abundance as well as declines of species, and even local extinctions [8,22,24]. Repeated surveys of only one or a few PAs provide useful insights but do not enable systematic comparisons of biodiversity changes across topoclimatic gradients.

Surprisingly, very few studies have specifically investigated the potential of topoclimatic buffering to support species persistence based on repeated surveys of multiple populations. Using a large set of climate-threatened species from the UK, Suggitt et al. [27] showed that fine-grained climatic variation caused by topographic heterogeneity may indeed buffer species against regional extirpations. These results are encouraging, but it should be noted that studying changes in species occurrences over coarse scale grid cells (10 km × 10 km) brings two challenges: separating potential land use-induced impacts on species populations (cf. [25,28]) from climate change impacts, and the potential omission of declines in species abundances which often become apparent earlier than changes in occupancies [29]. In another study, Maclean, et al. [30] examined the impacts of fine-resolution microclimatic buffering on grassland plant species responses and showed that cooler slopes can support the prolonged persistence of populations. What remains open to debate is how well these results apply to other species groups in other environments showing more rapid responses to climate change than plants which have many means to survive during unfavorable periods [31].

Clearly, we need more studies on changes in the abundance of species showing potentially rapid responses to climate change. It would be especially valuable to examine changes in species abundances systematically across PAs, with different potentials for topoclimatic buffering where human-induced land use changes are not allowed. Among different species groups, birds can be particularly useful in studying the impacts of topoclimatic variation. This is because bird populations have the potential to respond rapidly to the changes in climate, as shown by the recent changes in their distributions and in the composition of local bird communities [26,32]. Moreover, there are sufficient data on bird species abundances available from systematic bird surveys [29,33,34].

Here, we studied the temporal changes in bird species densities in 129 individual PAs with varying levels of topoclimatic heterogeneity, using systematically recorded data from a national PA network which may effectively alleviate the negative impacts of climate change on species [35–38]. We investigated the variation within the PAs in fine-grained air temperature patterns to assess whether larger temperature variation can slow down the negative effects of climate warming on bird species populations. We focused specifically on population size changes in northerly distributed bird species in a region where these species have contracting range-margin populations [39,40]. Alongside the temperature variability, we addressed two other potentially important drivers of bird population sizes: changes in the observed broad-scale air temperature (i.e., macroclimate) [41,42] and the size of the PAs [43,44], and also took the variation in forest proportion in the PAs into account.

Specifically, we addressed the following two questions: (1) Is the density of bird species populations higher in PAs showing a larger variation in local temperature (conditions)? (2) Is the density of northern bird species declining less in PAs with larger variation in local temperature (conditions)?

2. Material and Methods

2.1. Study Areas

Finland stretches 1100 km across the boreal biome of northern Europe (Figure 1). In this biome northern species show increasing and southern species decreasing trends towards the north. The boreal biome can be divided into southern, middle and northern boreal zones. Here, we focused on bird population trends in the middle boreal zone, where the ranges of southern and northern species largely overlap [45]. The PAs occurring in the middle boreal zone are mostly covered with coniferous (dominated by Scots pine *Pinus sylvestris* or Norway spruce *Picea abies*), mixed and deciduous (dominated mainly by birch *Betula* spp.) forests and pine and open, treeless mires.

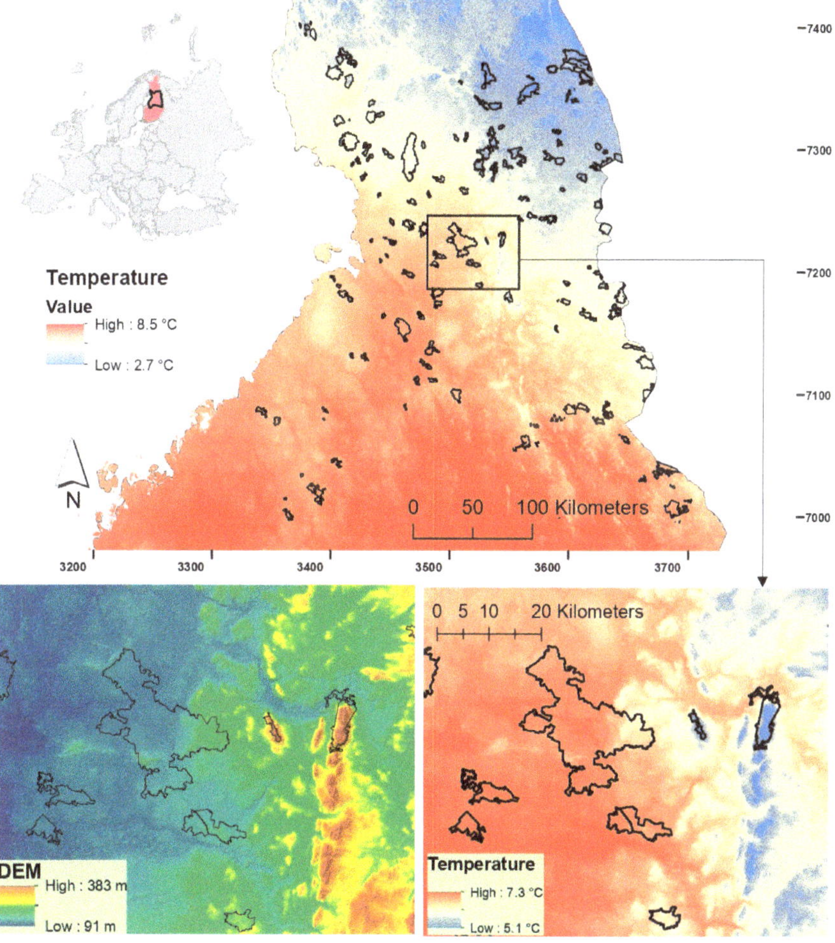

Figure 1. Location of the study areas (protected areas (PAs) in black lines) in Finland with the mean air temperature for April–May–June at 50 m × 50 m grid size. As an example, the topographic variation (meters above sea level; DEM = digital elevation model) and mean April–June air temperatures are shown for a specific region with a high variation in temperature caused by variation in topographic heterogeneity.

In this study, birds were counted in 129 PAs situated in the middle boreal zone both in 1981–1999 and in 2000–2017 (including the southernmost part of northern boreal zone; uniform coordinate system [epsg: 2393], grids 70–74, Figure 1). In these PAs the total land area was 4347 km^2, with a mean of 33.7 km^2 (size range 1.6–300.3 km^2, Figure 1, Table S1). The forests (including wooded mires) covered 63% of the land area of the studied PAs, the remaining being open, treeless mires. This information on the proportion of forests in the studied PAs is based on the data of habitat classification in National Parks Finland, which governs state-owned protected areas. More than two thirds of the protected forest stands are over 100 years old [46].

2.2. Bird Censuses

Land birds in the studied 129 PAs were counted using the Finnish line transect census method, which is suitable for counting birds over large areas [45,47]. The line transect method applies a one-visit census in which birds are counted during the breeding season along a transect with an average length of 5–6 km.

The census is carried out in late May and in June in the early morning, when the singing activity of the birds is highest. In the line transect method, a 50 m (25 m on each side)-wide main belt along the walking line and a supplementary belt outside the main belt are separated. Birds are counted both from the main belt and the supplementary belt which together constitute a survey belt. The supplementary belt consists of all birds observed outside the main belt (for details, see [45,47,48]).

In the Finnish line transect method, the densities of species based on the observations in the censuses are calculated in standard units of pairs/km^2. A pair is inferred (following the instructions for the Finnish line transect census) by: a male heard singing, an otherwise observed male or female, or a group of observed fledglings. Densities of bird species were calculated on the basis of observations in the whole survey belt. In calculating the density, a species-specific correction coefficient (based on the proportion of main belt observations in the whole survey belt) was used to take into account the differing audibility and other detectability of different species (see, e.g., [47,48]).

When counting the birds, in both time periods the total length of the transects in each of the 129 PAs was at least 1.0 km (see Figure 1, Table S1). The total length of the line transect censuses in the studied PAs amounted to 4677.8 km in 1981–1999, and 4585.9 km in 2000–2017, and the mean total length of the transects in a PA was 36.3 km in 1981–1999 and 35.5 km in 2000–2017 (Table S1). The median number of years when the censuses were carried out was two in each protected area, both in 1981–1999 and in 2000–2017. The median census year was 1994 in the first period and 2009 in the second period. Precisely the same transects were not surveyed, but the censuses in each protected area included the same proportion of habitats during the two periods.

The population changes of northern forest bird species revealed by the census data from the two time periods were related to the topoclimatic variation (50 m × 50 m) within the PA, the changes in the broad-scale (10 km × 10 km) climate, and the size and forest proportion of the PA. We concentrated only on forest species because open mire species—such as many waders—occur on flat peatlands and thus are inherently associated with breeding sites with a low variation in local climatic conditions. We included all bird species breeding in forests with a northern distribution in Finland. Based on these criteria, the following 17 species [49,50] were selected for our study: rough-legged buzzard *Buteo lagopus*, golden eagle *Aquila chrysaetos*, merlin *Falco columbarius*, northern hawk owl *Surnia ulula*, great grey owl *Strix nebulosa*, three-toed woodpecker *Picoides tridactylus*, Bohemian waxwing *Bombycilla garrulus*, red-flanked bluetail *Tarsiger cyanurus*, Arctic warbler *Phylloscopus borealis*, Siberian tit *Poecile cinctus*, great grey shrike *Lanius excubitor*, Siberian jay *Perisoreus infaustus*, brambling *Fringilla montifringilla*, common redpoll *Carduelis flammea*, two-barred crossbill *Loxia leucoptera*, pine grosbeak *Pinicola enucleator*, and rustic bunting *Emberiza rustica*. Over 80% of recorded bird pairs of the 17 species studied belong to the three most abundant species, i.e., the brambling, common redpoll and rustic bunting (see Table 1).

Table 1. Mean density (pairs/km^2 ± standard error) of northern forest bird species in 1981–1999 and in 2000–2017 in the 129 protected areas. The statistical significance between the time-periods is based on a paired Wilcoxon signed rank test with positive, negative and tied ranks from 1981–1999 to 2000–2017. Density values of 0.00 denotes density of less than 0.005 pairs/km^2.

Species	Density		Ranks (pos./neg./tied)	z	P
	1981–1999	2000–2017			
Rough-legged buzzard *Buteo lagopus*	0.01 ± 0.01	0.01 ± 0.00	1/6/122	−1.859	0.063
Golden eagle *Aquila chrysaetos*	0.00 ± 0.00	0.00 ± 0.00	10/6/113	1.189	0.235
Merlin *Falco columbarius*	0.04 ± 0.02	0.04 ± 0.02	10/14/105	−0.943	0.346
Northern hawk owl *Surnia ulula*	0.04 ± 0.02	0.28 ± 0.22	14/8/107	1.185	0.236
Great grey owl *Strix nebulosa*	0.01 ± 0.01	0.01 ± 0.01	2/6/121	−1.120	0.263
Three-toed woodpecker *Picoides tridactylus*	0.31 ± 0.05	0.65 ± 0.09	59/21/49	4.240	<0.001
Bohemian waxwing *Bombycilla garrulus*	0.21 ± 0.05	0.79 ± 0.09	82/7/40	6.880	<0.001
Red-flanked bluetail *Tarsiger cyanurus*	0.01 ± 0.00	0.14 ± 0.04	28/6/95	4.129	<0.001
Arctic warbler *Phylloscopus borealis*	0.01 ± 0.00	0.00 ± 0.00	3/7/119	−1.988	0.047
Siberian tit *Poecile cinctus*	0.14 ± 0.04	0.07 ± 0.03	8/17/104	−1.682	0.093
Great grey shrike *Lanius excubitor*	0.21 ± 0.09	0.07 ± 0.02	17/18/94	−0.541	0.589
Siberian jay *Perisoreus infaustus*	0.47 ± 0.07	0.43 ± 0.07	29/36/64	−0.709	0.478
Brambling *Fringilla montifringilla*	16.73 ± 0.97	9.95 ± 0.83	23/106/0	−7.229	<0.001
Common redpoll *Carduelis flammea*	3.03 ± 0.33	1.42 ± 0.23	19/100/10	−7.231	<0.001
Two-barred crossbill *Loxia leucoptera*	0.56 ± 0.14	0.03 ± 0.01	10/46/73	−4.968	<0.001
Pine grosbeak *Pinicola enucleator*	0.10 ± 0.04	0.06 ± 0.02	7/9/113	−0.310	0.756
Rustic bunting *Emberiza rustica*	3.43 ± 0.30	1.84 ± 0.28	32/82/15	−5.365	<0.001
Combined	25.29 ± 1.39	15.76 ± 1.23	20/109/0	−7.154	<0.001

2.3. Climate Data

Our climate data consisted of data recorded at two distinctly different spatial resolutions. First, as a measure to examine the effect of the broad-scale air temperature in the two time periods (i.e., 1981–1999 and 2000–2017), as well as the changes between the two periods, on the bird species densities, we used the mean air temperature values of April–June ($TAMJ_{broad}$) interpolated for each of the 10×10 km grid squares where a protected area was located. We focus on April–June air temperatures, because this is the time of arrival of migratory birds and the breeding period for all the studied northern bird species. The temperature values were obtained from the daily gridded climate dataset by the Finnish Meteorological Institute [51] both for the period of 1981–1999 and 2000–2017.

Second, in order to capture the variation in the local climatic conditions in the studied PAs, we modelled the monthly average air temperatures (1981–2010) across the study domain based on daily data from 318 meteorological stations derived from the European Climate Assessment and Dataset database (ECA&D; [52]). For these models, we employed generalized additive models (GAM), as implemented in R-package mgcv version 1.8-7 [53]. Following Aalto et al. [54], we used predictors of the geographical location, topography (elevation, potential radiation, relative elevation) and water cover [55] to produce monthly average air temperature surfaces at a high spatial resolution (50 m × 50 m) for all months. The modelled monthly average air temperatures agreed well with the observations from the meteorological stations, with a root mean squared error (RMSE; leave-one-out cross-validation) ranging from 0.56 (June; Figure S1) to 1.49 °C (January). We used the fine-grained (50 m) variation within the PAs ($\sigma TAMJ_{local}$, quantified as the standard deviation of April–June mean temperatures across PAs) in the subsequent analyses, as an indicator of local temperature variation driven by topographic heterogeneity, and thereby an indicator of potential buffering effect.

We could not account for the effect of forest canopy cover in our modeling of local temperatures. This is due to the placing of meteorological stations on the open landscape [56]. Although this is likely to increase modelling uncertainty in forested areas, the microclimatic effect of forest canopy (i.e., buffering temperature variability) is indirectly accounted for in the models by using the forest proportion predictor.

2.4. Statistical Analyses

We modeled the spatial variation in the population density of northern forest birds over 1981–2017 (i.e., our first response variable; combined density in 1981–1999 and in 2000–2017 in Table S1) as a function of the $TAMJ_{broad}$ and $\sigma TAMJ_{local}$. In addition, we used forest proportion and the size of the PA as predictors in our models. Large PAs are likely to have higher bird densities compared to small PAs [e.g., 44]. The size of the PA was log-transformed for the analyses to approximate normality.

In studying the bird density variation between the PAs, we used generalized linear mixed modelling (GLMM [57]; assuming negative-binomial errors with a log-link function), that accounts for zero-inflation in the response variable. In addition to the main effects of $\sigma TAMJ_{local}$ and forest proportion, we included their interaction with the model. This interaction was included because we expected that forest cover can mediate the effect of local temperature variability on bird densities. Indeed, $\sigma TAMJ_{local}$ and forest proportion were intercorrelated (Spearman's rank correlation $r_s = 0.706$). Species were considered as a random factor in the model, and the two time-periods were included as an ordered factor (1 = 1981–1999, 2 = 2000–2017). The model structure in R syntax:

$$\text{Bird density} = TAMJ_{broad} + \sigma TAMJ_{local} + \text{forest proportion} + \sigma TAMJ_{local} \times \text{forest proportion} + \log(\text{PA size}) + \text{time} + (1 \mid \text{species}) \tag{1}$$

Secondly, we modelled the change in bird species density from 1981–1999 to 2000–2017 (i.e., our second response variable) as a function of change in broad-scale air temperature ($\Delta TAMJ_{broad}$ change in April–June air temperatures between 2000–2017 and 1981–1999) between the two time periods, local temperature variability and size of the PA. In this model (GLMM), forest proportion in the PA was regarded as an offset variable because of the intercorrelation with $\sigma TAMJ_{local}$. The offset term is a structural predictor, the values of which are not estimated by the model but are assumed to have the value of one. The values of the offset are added to the linear predictor of the target. Here, we assumed Gaussian error distribution and included species as a random factor:

$$\text{Change in bird density} = \Delta TAMJ_{broad} + \sigma TAMJ_{local} + \log(\text{PA size}) + (1 \mid \text{species}) \tag{2}$$

The modeling was carried out in R (version 3.6.1 [58]) and the GLMMs were fitted using the package glmmTMB [57]. There was a slight positive correlation between $\Delta TAMJ_{broad}$ and $\sigma TAMJ_{local}$ ($r_s = 0.388$), $\sigma TAMJ_{local}$ and size of the PA ($r_s = 0.312$), and $\Delta TAMJ_{broad}$ and size of the PA ($r_s = 0.208$).

When comparing species-specific density changes in the PAs between the time periods, we used a non-parametric Wilcoxon signed rank test.

3. Results

We found that the combined density of the 17 studied northern forest bird species declined significantly ($p < 0.001$) by ca. 38% between 1981–1999 and 2000–2017 (Table 1). Five species—including all the three most abundant northern forest bird species (brambling, redpoll, rustic bunting)—and two less-abundant species (Arctic warbler, two-barred crossbill) declined significantly while three species increased (three-toed woodpecker, Bohemian waxwing, red-flanked bluetail).

Our modelling showed that broad-scale air temperature had a negative effect, and local temperature variability a positive effect, on the densities of northern forest bird species (Table 2). On average, densities of northern forest bird species increase northwards (Table S1). Therefore, the variation of $TAMJ_{broad}$ was negatively related to the bird abundances as values of April–June average air temperatures generally decline northwards. In addition, we found a significant ($p < 0.05$) negative interaction between forest and $\sigma TAMJ_{local}$, indicating that an increase in forest proportion inside PAs tends to decrease the effect of local temperature variability on bird densities.

Table 2. Results of generalized linear mixed modelling (GLMM; assuming negative-binomial errors with a log-link function) of bird population densities in PAs. Bird densities were modelled using broad-scale air temperature variation ($TAMJ_{broad}$), local air temperature variability ($\sigma TAMJ_{local}$), forest proportion, interaction between $\sigma TAMJ_{local}$ and forest proportion and PA size as predictors. Species were considered in the models as random factors. Predictor Time shows the division of data into two distinct parts. The protected area size was log-transformed prior to the analysis.

Term	Estimate	S.E.	z	P
Intercept	0.221	0.532	0.416	0.677
$TAMJ_{broad}$	−0.579	0.041	−14.232	<0.001
$\sigma TAMJ_{local}$	5.186	2.554	2.030	0.042
Forest proportion	1.485	0.201	7.382	<0.001
Forest proportion $\times$ $\sigma TAMJ_{local}$	−6.816	2.790	−2.443	0.0146
PA Size	0.013	0.030	0.417	0.677
Time	0.195	0.577	0.337	0.736

Local temperature variability had a significant ($p < 0.001$) negative effect on the density change of northern forest birds from 1981–1999 to 2000–2017 (Table 3). Importantly, the decline in bird density was smaller in PAs associated with large temperature variation. In contrast, change in broad-scale air temperatures ($\Delta TAMJ_{broad}$) did not have any significant effect on the density change of northern forest birds. The size of the PAs had a positive effect on the change of population densities of northern forest birds, so that densities declined more in large PAs.

Table 3. Results of generalized linear mixed modelling (GLMM; assuming Gaussian errors) of bird density change in PAs. Bird density was modelled using change in broad-scale air temperatures ($\Delta TAMJ_{broad}$), local air temperature variability ($\sigma TAMJ_{local}$) and PA size as predictors. Forest proportion was an offset variable in the analysis. The protected area size was log-transformed prior to the analysis.

Term	Estimate	S.E.	z	P
Intercept	−1.057	0.348	−3.041	0.002
$\Delta TAMJ_{broad}$	0.191	0.301	0.634	0.526
$\sigma TAMJ_{local}$	−3.426	0.465	−7.375	<0.001
PA size	0.078	0.035	2.213	0.027

4. Discussion

Our analyses show that local temperature variability had a positive effect on the population density of northern forest species in the time periods of 1981–1999 and 2000–2017. High temperature variation also seems to buffer the negative decline of northern forest birds. Our results are thus in line with earlier studies investigating the impacts of topoclimatic variation on the persistence of species populations that are based, as our study is, on data from repeated surveys [27,30,59].

Interestingly, broad-scale temperature affects the density patterns of northern forest birds, but change in temperature does not explain the decrease of these species in the model. This can be due to the fact that in the model local temperature variability outweighs the negative effects of climate warming on the northern declining bird species. Moreover, the range of change in broad-scale temperatures is much less than variation in local temperatures, so broad-scale temperatures probably affect more similarly everywhere. The negative effects of climate warming on the northern bird species have been observed in earlier studies with their ranges shifting northwards [50,60], population densities in PAs declining [61] and mean weighted densities moving northwards [47].

Thus, bird densities tend to be higher in topographically more heterogeneous PAs, and the local temperature variability seems to provide a buffering effect against the declining population trends of northern forest birds in our study area. High-latitude regions, such as our study region, have experienced more severe warming than other areas—on average twice the global rate of warming [62]. In our study region, the mean April–June temperature has increased by an average of 1.0 °C, and the mean annual temperature has increased by an average of 1.1 °C between the periods 1981–1999 and 2000–2017 [40]. Mirroring our results against these trends in climate, detection of workable mechanisms for retarding the negative effects of climate warming can be considered as of special importance.

Controlling for the potential land use-induced impacts on populations from climate change impacts in coarse-scale grids can be difficult, particularly as human interventions can cause both negative and positive impacts on species viability (e.g., loss vs. restoration of habitat). Previous studies have not been carried out in protected areas and, therefore, land-use change may bias the observed results, because in a flatter landscape land use is much more intensive than in a rough terrain or in mountains. In our PAs, human-caused habitat alteration is not allowed; therefore, our results are more explicitly related to the topographic heterogeneity and climate-induced changes than in earlier studies.

As regards the extend of vertical variation, our results are highly interesting. This is because topographic variation of our PAs is relatively moderate, with a maximum vertical within-PA difference of about 250 meters. Yet, this variation, together with differences in incoming solar radiation between North- and South-facing slopes [54], is sufficient to counteract the recent macroclimatic forcing impacts on bird populations. Thus, the buffering impacts of local air temperatures may not require very large elevational gradients but can be effective already in moderately varying landscape. A study by Gaüzère, Prince and Devictor [59] investigated the climatic debt in bird communities in France using the species community temperature index (CTI), and showed that an increasing range in elevation can affect bird species populations positively by reducing the rate of their change. However, the study area in France also included mountainous landscapes, where the range of elevational variation was measured in several hundreds of meters compared to our study PAs with the variation commonly measured in tens of meters.

A central difference between the present results and a study carried out in the UK [27] is that the latter was based on presence records recorded in Atlas data at a 10-km resolution, while our work examined species abundances and the buffering capacity of fine-grain temperature patterns at a 50 × 50 m resolution. In the UK study, the extirpation risks stemming from climate warming were reduced by 22% in plants and 9% in insects due to topographic variation creating potential for microrefugia [27]. However, assessing changes in species trends based on occurrences from such coarse-resolution Atlas data may only partly conceal the changes in species populations which would otherwise be visible in abundance trends [33,47,63]. Thus, by using abundance analyses, the effect of topographic heterogeneity might be even more clearly observable.

The benefits of topographically more heterogeneous PAs may also manifest themselves in single species ecological and behavioral phenomena; for example, by helping a species to avoid misusing phenological triggers and the associated temporal mismatches between trophic levels [64]. In the boreal region, in topographically heterogeneous southern Norway, the breeding success of black grouse *Tetrao tetrix* and the capercaillie *Tetrao urogallus* has been observed to increase under the warming climate [65]. In contrast, in the flatter boreal forests of Finland, populations of black grouse have been declining—which has been attributed to the mismatch between the advanced time of mating and chicks hatching too early, resulting in declining breeding success [66].

Our findings suggest that topoclimatic variation may provide an effective conservation measure against climate change in moderately rugged landscapes. Such areas can be common in many places in the western boreal Palaearctic, where landscape is covered by a peneplain with a moderately rough land surface with small-scale variability in topography produced by a long period of erosion and other physical processes [67]. In these vast regions in the boreal biome, topographical heterogeneity may provide a significant buffer against rapid climate-induced changes in many areas [19].

The results reported here also have important consequences for climate-smart conservation planning for boreal regions. In addition to support for protecting particularly topographically heterogeneous large areas for declining northern species, another tentative consideration and a potentially useful tactic in climate-smart conservation planning might be targeting for a sufficiently connected network of topoclimatically heterogeneous PAs. This kind of approach could provide climatically suitable stepping stones and holdouts for a population to persist for a limited period of time, thereby facilitating the range shifts of contracting species under deteriorating climatic conditions (cf. [20]).

5. Conclusions

Our study was carried out in protected areas where harmful land use, such as logging, is not allowed; therefore, our results of the positive effects of topographical heterogeneity on the populations of declining northern birds are rather straightforward. In contrast, in comparisons of lower topographical variability (e.g., flatlands) and higher topographic variability (rough terrain, mountains), land-use patterns may differ considerably and, thus, the lower land-use intensity may affect the buffering capacity of rough terrains against climate change. Further studies are also needed to demonstrate the significance of topographical heterogeneity in preserving biodiversity in a rapidly warming boreal climate outside protected areas with varying intensities of land use.

Supplementary Materials: The following are available online at http://www.mdpi.com/1424-2818/12/2/56/s1, Table S1. Protected areas studied. Figure S1. The agreement between observed and interpolated monthly average air temperatures (1981–2010).

Author Contributions: Conceptualization, J.A., M.L., R.K.H and R.V.; methodology, A.R., J.A., M.L., N.L., R.K.H. and R.V.; data analysis, J.A., R.H.K., M.L., N.L., R.V.; data curation, A.R., J.A., N.L., R.V.; writing—original draft preparation, A.R., J.A., M.L., N.L., R.K.H, R.V. and S.K.; project administration and funding acquisition, R.K.H., R.V. and S.K. All authors have read and agreed to the published version of the manuscript.

Funding: The study was part of a project evaluating the protected area network in the changing climate (SUMI) which was funded by the Ministry of the Environment in Finland and was financially supported also by the Strategic Research Council (SRC) at the Academy of Finland (IBC-Carbon, no 312559).

Acknowledgments: Bird censuses were carried out by numerous field ornithologists, whom we gratefully acknowledge. Most of the census data have been collected in Metsähallitus, National Parks Finland.

Conflicts of Interest: The authors declare no conflict of interest. The funders had no role in the design of the study; in the collection, analyses, or interpretation of data; in the writing of the manuscript; or in the decision to publish the results.

References

1. Pereira, H.M.; Leadley, P.W.; Proenca, V.; Alkemade, R.; Scharlemann, J.P.W.; Fernandez-Manjarrés, J.F.; Araújo, M.B.; Balvanera, P.; Biggs, R.; Cheung, W.W.L.; et al. Scenarios for Global Biodiversity in the 21st Century. *Science* **2010**, *330*, 1496–1501. [CrossRef]

2. Bellard, C.; Bertelsmeier, C.; Leadley, P.; Thuiller, W.; Courchamp, F. Impacts of climate change on the future of biodiversity. *Ecol. Lett.* **2012**, *15*, 365–377. [CrossRef]

3. Chen, I.-C.; Hill, J.K.; Ohlemüller, R.; Roy, D.B.; Thomas, C.D. Rapid range shifts of species associated with high levels of climate warming. *Science* **2011**, *333*, 1024–1026. [CrossRef] [PubMed]

4. Stephens, P.A.; Mason, L.R.; Green, R.E.; Gregory, R.D.; Sauer, J.R.; Alison, J.; Aunins, A.; Brotons, L.; Butchart, S.H.M.; Campedelli, T.; et al. Consistent response of bird populations to climate change on two continents. *Science* **2016**, *352*, 84–87. [CrossRef] [PubMed]

5. Spooner, F.E.B.; Pearson, R.G.; Freeman, R. Rapid warming is associated with population decline among terrestrial birds and mammals globally. *Glob. Chang. Biol.* **2018**, *24*, 4521–4531. [CrossRef] [PubMed]

6. Bowler, D.E.; Hof, C.; Haase, P.; Kröncke, I.; Schweiger, O.; Adrian, R.; Baert, L.; Bauer, H.-G.; Blick, T.; Brooker, R.W.; et al. Cross-realm assessment of climate change impacts on species' abundance trends. *Nat. Ecol. Evol.* **2017**, *1*, 0067. [CrossRef]

7. Auer, S.K.; King, D.I. Ecological and life-history traits explain recent boundary shifts in elevation and latitude of western North American songbirds. *Glob. Ecol. Biogeogr.* **2014**, *23*, 867–875. [CrossRef]

8. Tingley, M.W.; Koo, M.S.; Moritz, C.; Rush, A.C.; Beissinger, S.R. The push and pull of climate change causes heterogeneous shifts in avian elevational ranges. *Glob. Chang. Biol.* **2012**, *18*, 3279–3290. [CrossRef]

9. Freeman, B.G.; Freeman, A.M.C. Rapid upslope shifts in New Guinean birds illustrate strong distributional responses of tropical montane species to global warming. *Proc. Natl. Acad. Sci. USA* **2014**, *111*, 4490–4494. [CrossRef]

10. Forero-Medina, G.; Terborgh, J.; Socolar, S.J.; Pimm, S.L. Elevational Ranges of Birds on a Tropical Montane Gradient Lag behind Warming Temperatures. *PLoS ONE* **2011**, *6*, e28535. [CrossRef]

11. Barbet-Massin, M.; Thuiller, W.; Jiguet, F. The fate of European breeding birds under climate, land-use and dispersal scenarios. *Glob. Chang. Biol.* **2012**, *18*, 881–890. [CrossRef]

12. Corlett, R.T.; Westcott, D.A. Will plant movements keep up with climate change? *Trends Ecol. Evol.* **2013**, *28*, 482–488. [CrossRef] [PubMed]

13. Burrows, M.T.; Schoeman, D.S.; Richardson, A.J.; Molinos, J.G.; Hoffmann, A.; Buckley, L.B.; Moore, P.J.; Brown, C.J.; Bruno, J.F.; Duarte, C.M.; et al. Geographical limits to species-range shifts are suggested by climate velocity. *Nature* **2014**, *507*, 492–495. [CrossRef] [PubMed]

14. Ordonez, A.; Martinuzzi, S.; Radelo, V.C.; Williams, J.W. Combined speeds of climate and land-use change of the conterminous US until 2050. *Nat. Clim. Chang.* **2014**, *4*, 811–816. [CrossRef]

15. Tingley, M.W.; Darling, E.S.; Wilcove, D.S. Fine- and coarse-filter conservation strategies in a time of climate change. In *The Year in Ecology and Conservation Biology*; Ostfeld, R.S., Power, A.G., Eds.; The New York Academy of Sciences: New York, NY, USA, 2014; Volume 1322, pp. 92–109.

16. Nadeau, C.P.; Fuller, A.K.; Rosenblatt, D.L. Climate-smart management of biodiversity. *Ecosphere* **2015**, *6*, 91. [CrossRef]

17. Ackerly, D.D.; Loarie, S.R.; Cornwell, W.K.; Weiss, S.B.; Hamilton, H.; Branciforte, R.; Kraft, N.J.B. The geography of climate change: Implications for conservation biogeography. *Divers. Distrib.* **2010**, *16*, 476–487. [CrossRef]

18. Dobrowski, S.Z. A climatic basis for microrefugia: The influence of terrain on climate. *Glob. Chang. Biol.* **2011**, *17*, 1022–1035. [CrossRef]

19. Greiser, C.; Ehrlén, J.; Meineri, E.; Hylander, K. Hiding from the climate: Characterizing microrefugia for boreal forest understory species. *Glob. Chang. Biol.* **2019**, in press. [CrossRef]

20. Hannah, L.; Flint, L.; Syphard, A.D.; Moritz, M.A.; Buckley, L.B.; McCullough, I.M. Fine-grain modeling of species' response to climate change: Holdouts, stepping-stones, and microrefugia. *Trends Ecol. Evol.* **2014**, *29*, 390–397. [CrossRef]

21. Lawler, J.J.; Ackerly, D.D.; Albano, C.M.; Anderson, M.G.; Dobrowski, S.Z.; Gill, J.L.; Heller, N.E.; Pressey, R.L.; Sanderson, E.W.; Weiss, S.B. The theory behind, and the challenges of, conserving nature's stage in a time of rapid change. *Conserv. Biol.* **2015**, *29*, 618–629. [CrossRef]

22. Thomas, C.D.; Gillingham, P.K. The performance of protected areas for biodiversity under climate change. *Biol. J. Linn. Soc.* **2015**, *115*, 718–730. [CrossRef]

23. Lenoir, J.; Hattab, T.; Pierre, G. Climatic microrefugia under anthropogenic climate change: Implications for species redistribution. *Ecography* **2017**, *40*, 253–266. [CrossRef]

24. Rapacciuolo, G.; Maher, S.P.; Schneider, A.C.; Hammond, T.T.; Jabis, M.D.; Walsh, R.E.; Iknayan, K.J.; Walden, G.K.; Oldfather, M.F.; Ackerly, D.D.; et al. Beyond a warming fingerprint: Individualistic biogeographic responses to heterogeneous climate change in California. *Glob. Chang. Biol.* **2014**, *20*, 2841–2855. [CrossRef] [PubMed]

25. Oliver, T.H.; Morecroft, M.D. Interactions between climate change and land use change on biodiversity: Attribution problems, risks, and opportunities. *Wiley Interdiscip. Rev. Clim.* **2014**, *5*, 317–335. [CrossRef]

26. Reside, A.E.; VanDerWal, J.J.; Kutt, A.S.; Perkins, G.C. Weather, Not Climate, Defines Distributions of Vagile Bird Species. *PLoS ONE* **2010**, *5*, e13569. [CrossRef] [PubMed]

27. Suggitt, A.J.; Wilson, R.J.; Isaac, N.J.B.; Beale, C.M.; Auffret, A.G.; August, T.; Bennie, J.J.; Crick, H.Q.P.; Duffield, S.; Fox, R.; et al. Extinction risk from climate change is reduced by microclimatic buffering. *Nat. Clim. Chang.* **2018**, *8*, 713–717. [CrossRef]

28. Newbold, T.; Hudson, L.N.; Hill, S.L.L.; Contu, S.; Lysenko, I.; Senior, R.A.; Borger, L.; Bennett, D.J.; Choimes, A.; Collen, B.; et al. Global effects of land use on local terrestrial biodiversity. *Nature* **2015**, *520*, 45–50. [CrossRef]

29. Bowler, D.E.; Haase, P.; Kröncke, I.; Tackenberg, O.; Bauer, H.G.; Brendel, C.; Brooker, R.W.; Gerisch, M.; Henle, K.; Hickler, T.; et al. A cross-taxon analysis of the impact of climate change on abundance trends in central Europe. *Biol. Conserv.* **2015**, *187*, 41–50. [CrossRef]

30. Maclean, I.M.D.; Hopkins, J.J.; Bennie, J.; Lawson, C.R.; Wilson, R.J. Microclimates buffer the responses of plant communities to climate change. *Glob. Ecol. Biogeogr.* **2015**, *24*, 1340–1350. [CrossRef]

31. Freckleton, R.P.; Watkinson, A.R. Large-scale spatial dynamics of plants: Metapopulations, regional ensembles and patchy populations. *J. Ecol.* **2002**, *90*, 419–434. [CrossRef]
32. Lindström, Å.; Green, M.; Paulson, G.; Smith, H.G.; Devictor, V. Rapid changes in bird community composition at multiple temporal and spatial scales in response to recent climate change. *Ecography* **2013**, *36*, 313–322. [CrossRef]
33. Renwick, A.R.; Massimino, D.; Newson, S.E.; Chamberlain, D.E.; Pearce-Higgins, J.W.; Johnston, A. Modelling changes in species' abundance in response to projected climate change. *Divers. Distrib.* **2012**, *18*, 121–132. [CrossRef]
34. Howard, C.; Stephens, P.A.; Pearce-Higgins, J.W.; Gregory, R.D.; Willis, S.G. The drivers of avian abundance: Patterns in the relative importance of climate and land use. *Glob. Ecol. Biogeogr.* **2015**, *24*, 1249–1260. [CrossRef]
35. Virkkala, R.; Pöyry, J.; Heikkinen, R.K.; Lehikoinen, A.; Valkama, J. Protected areas alleviate climate change effects on northern bird species of conservation concern. *Ecol. Evol.* **2014**, *4*, 2991–3003. [CrossRef] [PubMed]
36. Gaüzère, P.; Jiguet, F.; Devictor, V. Can protected areas mitigate the impacts of climate change on bird's species and communities? *Divers. Distrib.* **2016**, *22*, 625–637. [CrossRef]
37. Santangeli, A.; Rajasärkkä, A.; Lehikoinen, A. Effects of high latitude protected areas on bird communities under rapid climate change. *Glob. Chang. Biol.* **2017**, *23*, 2241–2249. [CrossRef]
38. Lehikoinen, P.; Santangeli, A.; Jaatinen, K.; Rajasärkkä, A.; Lehikoinen, A. Protected areas act as a buffer against detrimental effects of climate change-Evidence from large-scale, long-term abundance data. *Glob. Chang. Biol.* **2019**, *25*, 304–313. [CrossRef]
39. Virkkala, R.; Rajasärkkä, A. Northward density shift of bird species in boreal protected areas due to climate change. *Boreal Environ. Res.* **2011**, *16*, 2–13.
40. Virkkala, R.; Rajasärkkä, A.; Heikkinen, R.K.; Kuusela, S.; Leikola, N.; Pöyry, J. Birds in boreal protected areas shift northwards in the warming climate but show different rates of population decline. *Biol. Conserv.* **2018**, *226*, 271–279. [CrossRef]
41. Loarie, S.R.; Duffy, P.B.; Hamilton, H.; Asner, G.P.; Field, C.B.; Ackerly, D.D. The velocity of climate change. *Nature* **2009**, *462*, 1052–1055. [CrossRef]
42. Garcia, R.A.; Cabeza, M.; Rahbek, C.; Araújo, M.B. Multiple Dimensions of Climate Change and Their Implications for Biodiversity. *Science* **2014**, *344*, 1247579. [CrossRef] [PubMed]
43. Boulinier, T.; Nichols, J.D.; Hines, J.E.; Sauer, J.R.; Flather, C.H.; Pollock, K.H. Higher temporal variability of forest breeding bird communities in fragmented landscapes. *Proc. Natl. Acad. Sci. USA* **1998**, *95*, 7497–7501. [CrossRef] [PubMed]
44. Virkkala, R.; Rajasärkkä, A. Spatial variation of bird species in landscapes dominated by old-growth forests in northern boreal Finland. *Biodivers. Conserv.* **2006**, *15*, 2143–2162. [CrossRef]
45. Väisänen, R.A.; Lammi, E.; Koskimies, P. *Distribution, Numbers and Population Changes of Finnish Breeding Birds*; Otava: Helsinki, Finland, 1998; p. 567. (In Finnish with an English Summary)
46. Virkkala, R.; Korhonen, K.T.; Haapanen, R.; Aapala, K. Protected forests and mires in forest and mire vegetation zones in Finland based on the 8th National Forest Inventory. *Finnish Environ.* **2000**, *395*, 1–49. (In Finnish with an English Summary)
47. Virkkala, R.; Lehikoinen, A. Patterns of climate-induced density shifts of species: Poleward shifts faster in northern boreal birds than in southern birds. *Glob. Chang. Biol.* **2014**, *20*, 2995–3003. [CrossRef] [PubMed]
48. Järvinen, O.; Väisänen, R.A. Correction coefficients for line transect censuses of breeding birds. *Ornis Fenn.* **1983**, *60*, 97–104.
49. Väisänen, R.A. Changes in the distribution of species between two bird atlas surveys: The analysis of covariance after controlling for research activity. *Ornis Fenn.* **1998**, *75*, 53–67.
50. Virkkala, R.; Lehikoinen, A. Birds on the move in the face of climate change: High species turnover in northern Europe. *Ecol. Evol.* **2017**, *7*, 8201–8209. [CrossRef]
51. Aalto, J.; Pirinen, P.; Jylhä, K. New gridded daily climatology of Finland: Permutation-based uncertainty estimates and temporal trends in climate. *JGR-A* **2016**, *121*, 3807–3823. [CrossRef]
52. Klok, E.J.; Tank, A. Updated and extended European dataset of daily climate observations. *Int. J. Clim.* **2009**, *29*, 1182–1191. [CrossRef]
53. Wood, S.N. Fast stable restricted maximum likelihood and marginal likelihood estimation of semiparametric generalized linear models. *J. R. Stat. Soc. Ser.* **2011**, *73*, 3–36. [CrossRef]

54. Aalto, J.; Riihimäki, H.; Meineri, E.; Hylander, K.; Luoto, M. Revealing topoclimatic heterogeneity using meteorological station data. *Int. J. Clim.* **2017**, *37*, 544–556. [CrossRef]

55. Aalto, J.; le Roux, P.C.; Luoto, M. The meso-scale drivers of temperature extremes in high-latitude Fennoscandia. *Clim. Dyn.* **2014**, *42*, 237–252. [CrossRef]

56. De Frenne, P.; Verheyen, K. Weather stations lack forest data. *Science* **2016**, *351*, 234. [CrossRef] [PubMed]

57. Brooks, M.E.; Kristensen, K.; van Benthem, K.J.; Magnusson, A.; Berg, C.W.; Nielsen, A.; Skaug, H.J.; Machler, M.; Bolker, B.M. glmmTMB Balances Speed and Flexibility Among Packages for Zero-inflated Generalized Linear Mixed Modeling. *R J.* **2017**, *9*, 378–400. [CrossRef]

58. R Development Core Team. *A Language and Environment for Statistical Computing*; R Foundation for Statistical Computing: Vienna, Austria, 2018.

59. Gaüzère, P.; Prince, K.; Devictor, V. Where do they go? The effects of topography and habitat diversity on reducing climatic debt in birds. *Glob. Chang. Biol.* **2017**, *23*, 2218–2229. [CrossRef]

60. Brommer, J.E. The range margins of northern birds shift polewards. *Ann. Zool. Fenn.* **2004**, *41*, 391–397.

61. Virkkala, R.; Rajasärkkä, A. Climate change affects populations of northern birds in boreal protected areas. *Biol. Lett.* **2011**, *7*, 395–398. [CrossRef]

62. Ruosteenoja, K.; Jylhä, K.; Kämäräinen, M. Climate projections for Finland under the RCP forcing scenarios. *Geophysica* **2016**, *51*, 17–50.

63. Howard, C.; Stephens, P.A.; Pearce-Higgins, J.W.; Gregory, R.D.; Willis, S.G. Improving species distribution models: The value of data on abundance. *Methods Ecol.* **2014**, *5*, 506–513. [CrossRef]

64. Crick, H.Q.P. The impact of climate change on birds. *Ibis* **2004**, *146*, 48–56. [CrossRef]

65. Wegge, P.; Rolstad, J. Climate change and bird reproduction: Warmer springs benefit breeding success in boreal forest grouse. *Proc. R. Soc. B Biol. Sci.* **2017**, *284*, 20171528. [CrossRef] [PubMed]

66. Ludwig, G.X.; Alatalo, R.V.; Helle, P.; Lindén, H.; Lindström, J.; Siitari, H. Short- and long-term population dynamical consequences of asymmetric climate change in black grouse. *Proc. R. Soc. B Biol. Sci* **2006**, *273*, 2009–2016. [CrossRef] [PubMed]

67. Seppälä, M. *The Physical Geography of Fennoscandia*; Oxford University Press: Oxford, UK, 2005; p. 432.

Article

Prioritizing Areas for Land Conservation and Forest Management Planning for the Threatened Canada Warbler (*Cardellina canadensis*) in the Atlantic Northern Forest of Canada

Alana R. Westwood [1,*], **J. Daniel Lambert** [2], **Leonard R. Reitsma** [3] **and Diana Stralberg** [1]

[1] Boreal Avian Modelling Project, Department of Renewable Resources, University of Alberta, 751 General Services Building, AB T6G 2H1, Canada; stralber@ualberta.ca
[2] Northern Woodlands, P.O. Box 270, Lyme, NH 03768, USA; dan@northernwoodlands.org
[3] Department of Biological Sciences, Plymouth State University, 17 High Street, Plymouth, NH 03264, USA; leonr@plymouth.edu
* Correspondence: alana.westwood@hotmail.com

Received: 14 December 2019; Accepted: 30 January 2020; Published: 4 February 2020

Abstract: Populations of Canada Warbler (*Cardellina canadensis*) are declining in Canada's Atlantic Northern Forest. Land conservancies and government agencies are interested in identifying areas to protect populations, while some timber companies wish to manage forests to minimize impacts on Canada Warbler and potentially create future habitat. We developed seven conservation planning scenarios using Zonation software to prioritize candidate areas for permanent land conservation (4 scenarios) or responsible forest management (minimizing species removal during forest harvesting while promoting colonization of regenerated forest; 3 scenarios). Factors used to prioritize areas included Canada Warbler population density, connectivity to protected areas, future climate suitability, anthropogenic disturbance, and recent Canada Warbler observations. We analyzed each scenario for three estimates of natal dispersal distance (5, 10, and 50 km). We found that scenarios assuming large dispersal distances prioritized a few large hotspots, while low dispersal distance scenarios prioritized smaller, broadly distributed areas. For all scenarios, efficiency (proportion of current Canada Warbler population retained per unit area) declined with higher dispersal distance estimates and inclusion of climate change effects in the scenario. Using low dispersal distance scenarios in decision-making offers a more conservative approach to maintaining this species at risk. Given the differences among the scenarios, we encourage conservation planners to evaluate the reliability of dispersal estimates, the influence of habitat connectivity, and future climate suitability when prioritizing areas for conservation.

Keywords: bird distribution and abundance; boreal birds; Canada Warbler; *Cardellina canadensis*; Zonation; reserve design

---

1. Introduction

Canada Warbler (*Cardellina canadensis*) is a Neotropical migratory songbird that breeds in forests of the eastern U.S. and across Canadian forests from Nova Scotia to the Yukon [1]. Due to ongoing population declines (71% decline reported from 1970–2010 [2]), it is listed as Threatened in Canada [3] and a Species of Greatest Conservation Need in nearly every U.S. state where it breeds (e.g., [4]). The Canada Warbler International Conservation Initiative (CWICI) has urged research and conservation actions to help reverse the population decline, including identification of suitable habitat and development of best management practices for the breeding grounds [5].

In the Atlantic Northern Forest (Bird Conservation Region (BCR) 14), Canada Warbler population declines from 1970–2017 have been much steeper than those observed in other BCRs by the Canadian Breeding Bird Survey [6]. Canada Warblers predominantly use wet forests [7–10] and post-harvest deciduous and mixed-wood forests approximately 10–30 years of age [11,12]. Management guidelines recently produced for BCR 14 [13] (see Supplementary Materials) describe two different approaches to promote recovery of this species: permanent land conservation (hereafter 'land conservation', or 'LC') and responsible forest management (hereafter 'forest management', or 'FM'). These approaches were developed in tandem with, and designed for, two different user communities with distinct needs: (1) land conservancies and government agencies with a mandate to protect species and habitat in situ; and (2) forest industry staff and government agencies with a mandate to engage in sustainable development of forest resources while minimizing impacts to migratory birds and species at risk. LC scenarios (summarized in Westwood et al. [13]) (see Supplementary Materials) are intended to support conservation activities such as in-situ permanent protection of forested areas supporting Canada Warbler populations, forest blocks with low edge-to-interior ratios, and suitable habitat patches connected by forested corridors to other potential breeding sites to facilitate dispersal. FM scenarios are intended to support responsible forest management activities such as providing a continuous current supply of breeding habitat on the landscape, avoiding harvesting and road building in population centers and forested wetlands, and locating areas to implement silvicultural systems (such as shelterwood cutting) most likely to produce desired conditions for breeding habitat 12–20 years post-harvest, among other strategies [13] (see Supplementary Materials).

Forest fragmentation has been postulated as a cause of population declines for this species [14,15], so habitat connectivity on the breeding grounds is likely to play an important role in population recovery, especially given the high degree of conspecific attraction that has been documented [16]. Climate change is another important consideration for Canada Warbler, as its range is expected to retract under warming conditions [17]. Conservation planning software makes it possible to account for connectivity, climate change, and other factors when prioritizing areas for land conservation and forest management [18–20].

Most analyses using conservation planning software are designed to support the planning or evaluation of protected area networks (e.g., [21–23]). Other objectives include locating zones of interest for further study [24], evaluating trade-offs between land uses [25,26], or estimating the value of ecosystem services [27]. Conservation planning algorithms can be informed by estimates of dispersal, which is an important factor in wildlife habitat selection and response to climate change (i.e., the ability to move from current to future suitable habitats). For the Canada Warbler, published estimates of natal dispersal (distance from an individual's birth site to their breeding site) are not available. Estimates of natal dispersal for other passerines are frequently based on small sample sizes with a great deal of uncertainty, and estimates for species of similar sizes to the Canada Warbler range widely [28–31]. Conservation planning exercises are typically based on single dispersal estimates, and the implications of uncertainty in these estimates are unknown. Furthermore, conservation planning algorithms often rely on species distribution models which predict habitat suitability, species occurrence, or population density across a landscape. However, these models are inherently more uncertain in under-sampled areas, and this uncertainty is also not always accounted for during conservation prioritization exercises.

Conservation prioritization exercises are most useful when they directly inform resource allocation decisions across landscapes [32]. Therefore, we consulted a variety of government, conservation, and forest industry stakeholders to develop regional scenarios for prioritization [33] (see Supplementary Materials). We then used the program Zonation [34] to evaluate priority areas for land conservation and forest management to support Canada Warbler populations in the Canadian portion of BCR 14. One set of scenarios was designed to prioritize areas for long-term land conservation (including future climate change and connectivity to protected areas) and another set of scenarios prioritized areas for forest management on tenured land managed by timber companies.

We evaluated three estimates of natal dispersal distance for each set of scenarios. We evaluated the impact of these factors on spatial outcomes and identified areas that were consistently prioritized. We also compared conservation efficiency (proportion of the species' current estimated population protected per unit area) across scenario types and dispersal distance estimates. The resulting rankings of the landscape in BCR14 are intended to support decisions for maintaining and managing Canada Warbler habitat in this region by the two different user groups already described.

2. Materials and Methods

2.1. Study Area

The North American Bird Conservation Initiative (NABCI) defines BCRs as ecologically distinct regions that share similar bird communities, habitats, and resource management issues [35]. We selected the scale of the study as the Canadian portion of the Atlantic Northern Forest (BCR 14) for three reasons. First, BCRs are used by Canadian government agencies as planning and management units for other bird species at risk (e.g., [24,36]). Secondly, the Canada Warbler shows evidence of differential habitat selection across BCRs [37], and we wished to limit the modelling to a population with shared habitat requirements. Finally, limiting the scale of the study to the Canadian portion of the BCR allowed us to use interoperable data and take advantage of existing Canadian bird conservation and forestry networks comprised of government agencies, industry, and NGOs.

The Canadian portion of the Atlantic Northern Forest encompasses 20.4 million ha, including Nova Scotia, Prince Edward Island, New Brunswick, and the Gaspé Peninsula of Québec ([38]; Figure 1). This forest is within the Appalachian-Acadian Ecoregion, which also includes much of Maine and Vermont, and parts of New York and New Hampshire. Human activities in this area have influenced 98.2% of the land area [39].

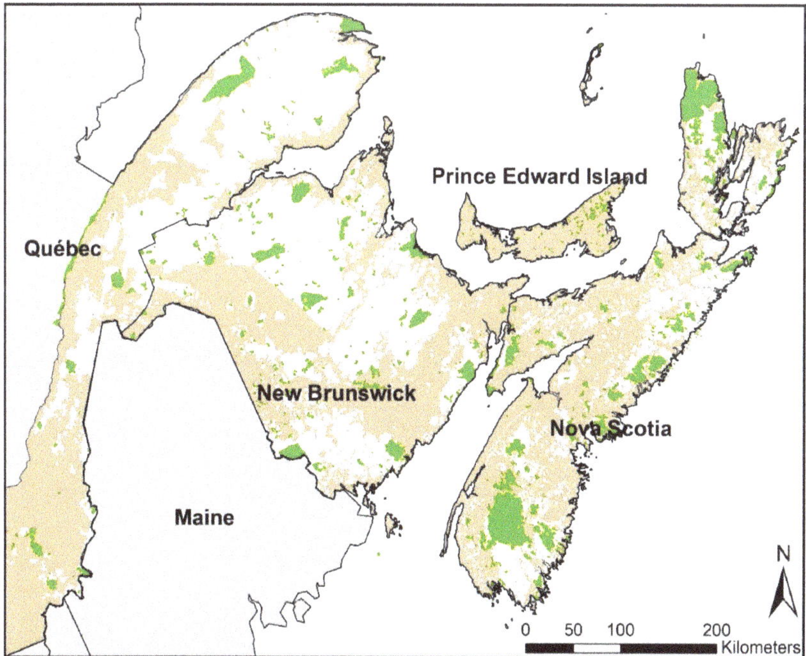

Figure 1. The Canadian portion of the Atlantic Northern Forest (Bird Conservation Region 14) showing protected areas (green) and areas under lease for timber activity (white).

The Atlantic Northern Forest is at the interface of two biomes, and includes tree species from the northern boreal forest (e.g., Black Spruce *Picea mariana*, Eastern White Pine *Pinus strobus*, Balsam Fir *Abies balsamea*) and southern temperate deciduous forest (e.g., Red Spruce *Picea rubens*, Red Maple *Acer rubrum*, Sugar Maple *Acer saccharum*). Land ownership is predominantly private, with the remainder under stewardship of the provinces. Known areas that are permanently protected from development in this region comprise 1.4 million ha (7% of the land area), including national and provincial parks, wilderness areas, and private protected landholdings [40]. Timberlands leased or owned by industrial forestry companies cover 6.7 million ha (33% of the land area). We did not include privately owned woodlots or timberlands in our analyses as we were unable to obtain a comprehensive land use map for such properties.

2.2. Scenario Development

Recent habitat guidelines for Canada Warbler in the Atlantic Northern Forest identified two types of stewardship approaches that could benefit this species: land conservation and forest management [13] (see Supplementary Materials). Based on solicited input from land trust representatives, agency and industrial forest managers, and scientists familiar with Canada Warbler ecology, we developed four scenarios for land conservation and three for forest management (Table 1; see [33] (see Supplementary Materials) for a description of stakeholder engagement and [13] (see Supplementary Materials) for specific details on recommended responsible forest management practices for maintaining Canada Warbler populations on the landscape).

Table 1. Spatial prioritization scenarios developed to support land conservation and forest management planning for Canada Warbler conservation in the Atlantic Northern Forest.

Scenario Group	Scenario Code	Description of Areas Prioritized	Included Data Layers
Land Conservation	LC1	Areas of high current Canada Warbler (CAWA) population density	*CAWA presence 2005–2009, CAWA presence 2010–2015, Population density, Model uncertainty*
	LC2	Areas of high current population density that are connected to protected areas	*CAWA presence 2005–2009, CAWA presence 2010–2015, Population density, Model uncertainty, Protected areas*
	LC3	Areas of high predicted population density under climate change that are connected to areas of high current population density	*CAWA presence 2005–2009, CAWA presence 2010–2015, Population density, Model uncertainty, CAWA climate baseline, CAWA climate 2050*
	LC4	Areas of high predicted population density under climate change that are connected to areas of high current population density and protected areas	*CAWA presence 2005–2009, CAWA presence 2010–2015, Population density, Model uncertainty, Protected areas, CAWA climate baseline, CAWA climate 2050*
Forest Management	FM1	Areas to serve as buffers or reserves during timber harvesting that have high population density and occur in wet forests (assumed to have low timber productivity)	*CAWA presence 2005–2009, CAWA presence 2010–2015, Population density, Model uncertainty, Working lands, Wet-poor habitat*
	FM2	Areas to harvest in upland forest areas (assumed to have high timber productivity) connected to wet forest areas with high Canada Warbler density to encourage colonization at 10–30 years post-harvest	*Population density, Model uncertainty, Working lands, Wet-poor habitat* *Population density, Model uncertainty, Working lands, Wet-poor habitat, Upland habitat, Protected areas*
	FM3	Areas to harvest in upland forest areas (assumed to have high timber productivity) with active avoidance of areas of high population density	*Subtraction of results of FM1 from FM2*

2.3. Zonation Spatial Conservation Prioritization

2.3.1. Data layers and Pre-Processing

To develop the seven scenarios, we acquired input data in three categories: avian, landcover, and administrative boundaries (Table 2).

Table 2. Description of data layers used to generate prioritization scenarios for Canada Warbler in the Atlantic Northern Forest (Bird Conservation Region 14).

Category	Spatial Data Layer	Description	Year Pub-lish-ed	Units	Data Ownership
Administrative Boundaries	Protected areas	The Conservation Areas Reporting and Tracking System geodatabase contains data on national, provincial, and privately-held protected lands, updated on an annual basis.	2015	categories	Canadian Council on Ecological Areas
	BCR14	Canadian portion of Bird Conservation Region 14—Atlantic Northern Forest.	2013	categories	NABCI
	Provinces/states	Federal, provincial, and state administrative regions.	2000	categories	ESRI 2000
	Working lands	Extent of actively leased or owned tenures by forestry companies, including both private and Crown lands.	2014	categories	Global Forest Watch Canada; Environment Canada
	Public lands	Unprotected public hands held by the Crown or other government bodies. May be currently under lease.	2013–2016	categories	Government (Gov't) of Nova Scotia (NS); Gov't of New Brunswick (NB), Gouv't du Québec, Gov't of Prince Edward Island (PEI)
Avian Data	Population density	Mean predicted population density of Canada Warbler in 2014 based predominantly on landcover and disturbance variables.	2013	males/hectare	Boreal Avian Modelling (BAM) Project, Haché et al. 2014
	Model uncertainty	Standard deviation of population density of Canada Warbler in 2014 based predominantly on landcover and disturbance variables.	2013	males/hectare	BAM, Haché et al. 2014
	CAWA presence 2005–2009	Locations where Canada Warblers were observed in point count surveys between 2005–2009.	2015	presence	BAM
	CAWA presence 2010–2015	Locations where Canada Warblers were observed in point count surveys between 2010–2015.	2015	presence	BAM
	CAWA climate baseline	Mean projected population density of Canada Warbler from 1961–1990.	2014	males/hectare	BAM, Stralberg et al. 2015
	CAWA climate 2050	Mean projected population density of Canada Warbler in 2050 based predominantly on climate-related variables.	2014	males/hectare	BAM, Stralberg et al. 2015
Landcover	Water bodies	Provincial hydrographic features at 1:10,000 scale.	2008–2014	N/A	NS Department of Natural Resources; NB Department of Natural Resources/GeoNB (link); Gov't of Québec (pers. comm.); Gov't of PEI;
	Wet-poor habitat	Treed areas categorized as wet-poor by the Northeastern Habitat Types Classification developed for ecosystems and habitats in the Northeast US and Atlantic Canada.	2015	N/A	The Nature Conservancy—Eastern Conservation Science; Ferree and Anderson 201
	Upland habitat	Treed areas categorized as wet-poor by the Northeastern Habitat Types Classification developed for ecosystems and habitats in the Northeast US and Atlantic Canada.	2015	N/A	The Nature Conservancy—Eastern Conservation Science; Ferree and Anderson 2013

Our avian datasets were obtained from the Boreal Avian Modelling (BAM) Project [41–43]. The BAM project holds the largest dataset of boreal and hemiboreal observations of birds in North America and accounts for heterogeneity in survey protocols by correcting abundance estimates for detectability [42]. To determine suspected breeding locations, we divided point occurrences of Canada Warblers into two groups (*CAWA presence 2005–2009* and *CAWA presence 2010–2015*) to capture areas showing persistent observations of Canada Warblers over time. Including recent presence information allows conservation planners to locate areas of high likelihood of extant Canada Warblers to consider for protection, and forest managers to avoid or operations in areas where they may harm, kill, or harass Canada Warbler or their nests or eggs (which are illegal activities under Canada's *Species At Risk Act, S.C. 2002*). To predict *Population density* in a 1 km grid across the study area, we used a national-scale species distribution model for Canada Warbler based on landcover and disturbance data [44]. The standard deviation of the population density estimates across multiple data subsets was used to represent *Model uncertainty*. As Zonation uses an additive approach to discount cell values by uncertainty, we chose a measure of uncertainty in the same units as the population density estimates (standard deviation), rather than using a standardized measure such as coefficient of variation. We note that no data were available on productivity or occupancy for the Canada Warbler in this region, nor fine-scale data to describe habitat characteristics (e.g., Lidar).

To account for projected climate-induced shifts in abundance, we used 4 km population density predictions from models based on climate, land-use, and topography covariates as compared to the baseline population density (detailed description in [45]). *CAWA climate baseline* included predicted population density from 1961–1990. Future projections were for the 2041–2070 time period (*CAWA climate 2050*) based upon a high-end, business-as-usual emissions scenario (A2), averaged over an ensemble of four global climate models from the Coupled Model Intercomparison Project (CMIP3) dataset [46]. Although these future projections do not incorporate anticipated lags in vegetation responses to climate change [17], we considered them a reasonable representation of long-term future habitat suitability for this species, which in this region is projected to experience relatively moderate shifts in density, as opposed to wholesale range shifts [47].

Administrative boundary layers included political and administrative borders (*Provinces/states* and *BCR14*), known forestry tenures (*Working lands*: lands owned or leased by forestry companies for the purpose of industrial development; [47]), and *Protected areas* meeting any of the International Union for the Conservation of Nature protected areas classifications I-IV (lands protected for long-term biological values, [40]).

Landcover layers included *Water bodies* derived from 1:10,000 aerial imagery and layers of *Wet-poor habitat* and *Upland habitat* (derived from the Northeastern Habitat Types Mapping Initiative; [48]). All spatial data were processed using ArcGIS 10.2.2 [49]. All input and output data layers were rasters in geotiff format with a cell size of 1000 × 1000 m, projected in Canada Lambert Conformal Conic.

2.3.2. Prioritization Analysis

We ran all analyses using Zonation 4.0 [18] with a mask of the land boundaries of the Atlantic Northern Forest (an inverse of the layer *Water bodies*) applied to eliminate oceans and lakes. Detailed run settings and input files are available at https://github.com/borealbirds/cawa-bcr-14. Zonation identifies areas with high concentrations of features, which are the items that are desirable to prioritize for the end use (e.g., population density, land ownership, etc.). Functions are used to apply rules to determine how the features interact and how connectivity between the features are prioritized or penalized [34].

Connectivity functions in Zonation rely on estimates of dispersal distance to determine whether features or populations are 'connected.' Due to scarce species-specific dispersal data, Carroll et al. [50] completed a prioritization analysis using an estimated dispersal distance (not specific to natal or breeding) of 10 km for landbird species. For Canada Warbler, there are no published natal dispersal estimates and only one known direct observation (10 km; L. Reitsma, unpublished data). Because natal dispersal distances of landbirds are correlated with wing length and body mass [29,30], the Canada Warbler's average size (wing span = 20–22 cm, mass = 9.5–12.5 g; Reitsma et al. 2010) suggests a median dispersal distance of 50 km [30]. However, estimates for similar-sized species vary widely, from 0.5 km to 40 km [28–31]. Betts et al. [51] measured maximum breeding dispersal distances of 1–3 km for two species of forest-dwelling warblers whose ranges overlap that of Canada Warbler (the Black-throated Blue Warbler, *Setophaga caerulescens*, and the Blackburnian Warbler, *Setophaga fusca*). To account for the uncertainty regarding dispersal estimates for this species and capture the variation in dispersal estimates for similar species, we evaluated three different natal dispersal estimates for each scenario: low dispersal distance (LDD, 5 km), medium dispersal distance (MDD, 10 km), and high dispersal distance (HDD, 50 km).

Zonation uses three primary algorithms to prioritize raster cells for selection: core area zonation (CAZ), additive benefit function, and targets [18]. In this case we did not have an a priori conservation target, which is one reason we chose Zonation over the similar software Marxan [19], which requires proportional or area targets as inputs. We chose CAZ because it ensures that the most valuable areas for each feature ("core areas") are prioritized, rather than allowing trade-offs between features, as is the case with additive benefit functions. In each iteration of the algorithm, CAZ chooses the cell with the lowest retention value as specified by the features and connectivity functions, discards it, and then recalculates the value of all remaining cells. In this way, each cell is ranked in order of its priority for selection, with the first cells selected being lowest priority and the last cells being highest priority [18].

For scenario LC1, we input the *Population density* feature (weight 1.0). In Zonation, features are generally given a standard weight of 1.0 unless they are to be discounted due to uncertainty (such as future climate projections) or increased in value due to particular management objectives [18]. We then applied the 'Distribution Smoothing' function to aggregate areas with cells of high population density connected by dispersal ability of Canada Warbler (dispersal kernel specified by α [18]; LDD = 5 km, $\alpha = 4 \times 10^{-4}$; MDD = 10 km, $\alpha = 2 \times 10^{-4}$ HDD = 50 km, $\alpha = 4 \times 10^{-5}$). The 'Species of Special Interest' function was used to increase the value of cells overlapping with *CAWA presence 2005–2009* and/or *CAWA presence 2010–2015*. Finally, we subtracted calculated *Model uncertainty* from all cells. After applying all features and functions, cells were prioritized for selection by the CAZ algorithm. LC2 included all features and functions from LC1. We then added *Protected areas* as a feature (weight 1.0) and used 'Matrix Connectivity' to prioritize selection of cells containing high Canada Warbler densities within 5 km of protected areas (distance specified by applying weighting factor = 0.1). The 'Matrix Connectivity' function multiplies the value of a cell based on its connectivity to other specified features, with features being considered connected if they fall within the specified dispersal distance [18].

LC3 included all features and functions from LC1 with the addition of the 'Ecological Interactions' function, which prioritizes areas based on connectivity between a pair of features [18]. We thus prioritized areas where high densities of Canada Warbler in both *CAWA climate baseline* (weight 0.75) and *CAWA climate 2050* (weight 0.75) were connected based on dispersal distance over 36 years (dispersal ability kernel is represented by β [18]; LDD = 5 km dispersal/year, 180 km total, $\beta = 1.1 \times 10^{-5}$; MDD = 10 km dispersal/year, 360 km total, $\beta = 5.56 \times 10^{-6}$; HDD = 50 km dispersal/year, 1800 km total, $\beta = 1.1 \times 10^{-6}$). We weighted climate features at 0.75 because assumptions about the future are less certain than present estimates (Moilanen and Arponen 2011). LC4 included all features and functions from LC2, with the addition of the 'Ecological Interactions' function used in LC3, in order to prioritize areas connecting high current and future populations of Canada Warbler with each other and with *Protected areas*.

FM1 included all features and functions from LC1 and added the 'Administrative Units' function, which recognizes that conservation decisions can be limited by administrative boundaries [18]. *Working lands* were given a weight of 0.5 to prioritize these areas for selection while maintaining connectivity with outside areas. The 'Matrix Connectivity' function was used to prioritize areas connecting high population density and *Wet-poor habitat*. FM2 included all features and functions from LC1 while omitting the 'Species of Special Interest' function to avoid prioritizing cells with Canada Warbler occurrences. 'Matrix Connectivity' was added to prioritize areas connecting high population density and *Upland habitat* (weight = 0.5). This function is based on the assumption that upland areas have desirable timber value and a greater potential to support Canada Warbler populations in 10–30 years after harvest if connected to dispersal sources and managed appropriately [13] (see Supplementary Materials). We added a second 'Matrix Connectivity' function to disincentivize prioritization of areas connecting high population density with *Wet-poor habitat* (weight = −0.5), and a third 'Matrix Connectivity' function deterring prioritization of areas connecting high population density with *Protected areas* (weight = −0.5).

FM3 was not completed in Zonation, but rather was derived by subtracting the FM2 solution raster from the FM1 solution raster in ArcGIS. This was intended to identify areas for timber harvest with the lowest risk of harm to populations and maximum economic opportunity.

2.3.3. Scenario Comparisons

For scenarios LC1–4 and FM1–2, we plotted performance curves representing conservation efficiency (the proportion of current predicted Canada Warbler population protected as a function of area protected, or not harvested in the case of FM2, at each dispersal distance estimate). Because of the different objectives of the scenarios, and functions applied, it was not sensible to compare conservation efficiency of all scenarios. We compared efficiency curves for scenarios with and without climate change but with all other factors being equal (LC2 and LC4). We also compared FM1 and FM2. Finally, we compared mean differences in cell-level rankings for all land conservation scenarios and estimated dispersal distances.

3. Results

For high resolution maps and data for all scenarios, visit https://github.com/borealbirds/cawa-bcr-14. Of the land conservation scenarios (Figure 2), areas that were consistently prioritized in both current and future climate scenarios included lands in central New Brunswick and the southeastern part of Québec along the border with Maine. All land conservation scenarios prioritized cells with recent and repeated observations of Canada Warbler, but these cells had a minimal effect on conservation efficiency or overall distribution of priority areas (e.g., Figure 3). Relatively few areas in Nova Scotia were prioritized in both current and future climate scenarios, except for small areas close to two large national parks. Adding connectivity to protected areas had little impact on the geographic distribution of areas prioritized in the current climate scenario (LC2), but had a more dramatic effect in the future climate scenario (LC4).

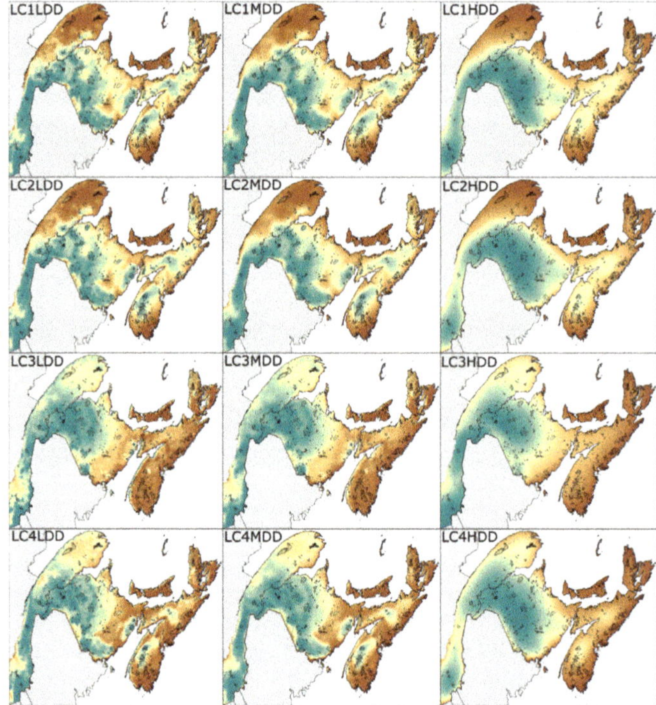

Figure 2. Zonation scenarios to prioritize areas for land conservation for Canada Warbler in Bird Conservation Region 14 at three natal dispersal distances: low dispersal distance (LDD, 5 km), medium dispersal distance (MDD, 10 km), and high dispersal distance (HDD, 50 km). Scenario descriptions given in Table 1. Priority rank for conservation scaled from blue (highest) to brown (lowest). Protected areas indicated by outlined polygons.

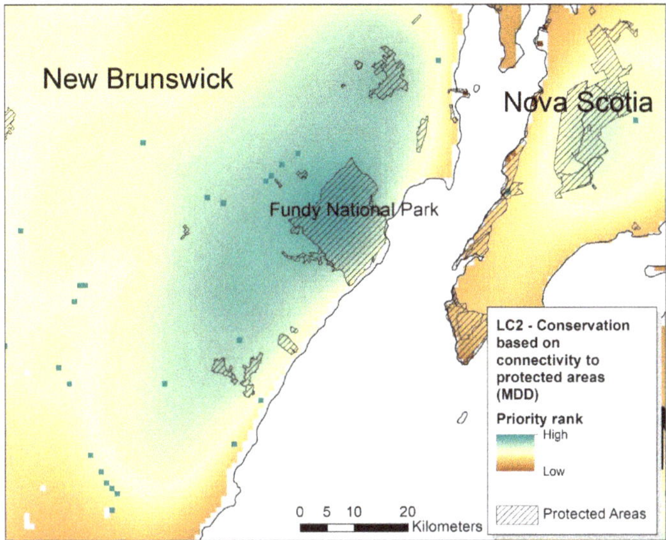

Figure 3. Zonation scenario to prioritize areas for land conservation with high current population density of Canada Warbler which are connected to protected areas at a medium natal dispersal distance (10 km). Crosshatched polygons indicate protected areas. Blue squares were highly prioritized because they included observations of Canada Warbler made during point count surveys between 2005 and 2015.

Increasing the dispersal distance estimate increased the aggregation of prioritized areas, with clusters of priority sites being larger and fewer in number for high dispersal distance scenarios, and smaller and more widely distributed for low dispersal distance scenarios (Figure 2). Overall, in the high dispersal distance scenarios, areas in Nova Scotia were rarely prioritized, with the largest aggregations of prioritized cells occurring in central New Brunswick, and to a lesser extent, southeastern Québec for current climate scenarios, and almost exclusively in Québec for future climate scenarios.

With the forest management scenarios (Figure 4), the emphasis was on forestry tenures; thus results are most appropriately applied by individual managers to compare the relative values of areas within those tenures when considering where to engage in responsible forest management practices (guidance is available in an accompanying technical report [33]) (see Supplementary Materials). Both FM1 and FM2 were affected by dispersal distance, shifting from many prioritized areas in Québec under LDD and MDD scenarios to prioritizing almost exclusively areas in New Brunswick under HDD scenarios.

When subtracting retention areas (FM1) from harvest priorities (FM2), the resulting scenario FM3 indicated that areas in the northern Gaspé Peninsula of Québec would be most effective at maximizing harvest value while minimizing potential loss of Canada Warbler. As with the land conservation scenarios, the prioritized areas were more aggregated with increased dispersal distance.

Scenario Comparison

For all scenarios, efficiency (proportion of current predicted Canada Warbler population retained per unit area) declined with higher dispersal distance estimates. Including climate change reduced land conservation scenario efficiency with respect to the current population (Figure 5). Efficiency was not directly comparable between forest management and land conservation scenarios due to differing objectives, as evidenced by the inverted efficiency curves for FM1 and FM2 (Figure 5), the former of which was designed to prioritize areas to avoid during forest management with the latter prioritizing areas to target for harvesting.

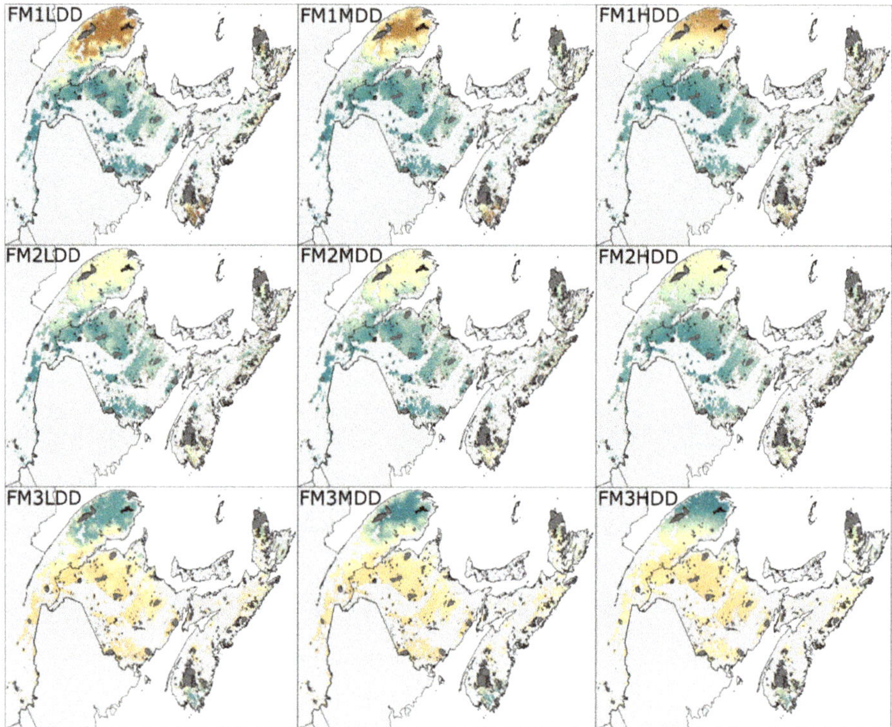

Figure 4. Zonation scenarios to prioritize areas for responsible forest management for Canada Warbler within forestry tenures in Bird Conservation Region 14 at three natal dispersal distances: 5 km (LDD), 10 km (MDD), and 50 km (HDD). Scenario descriptions given in Table 1. Priority rank for management scaled from blue (highest) to brown (lowest). Protected areas indicated by gray polygons.

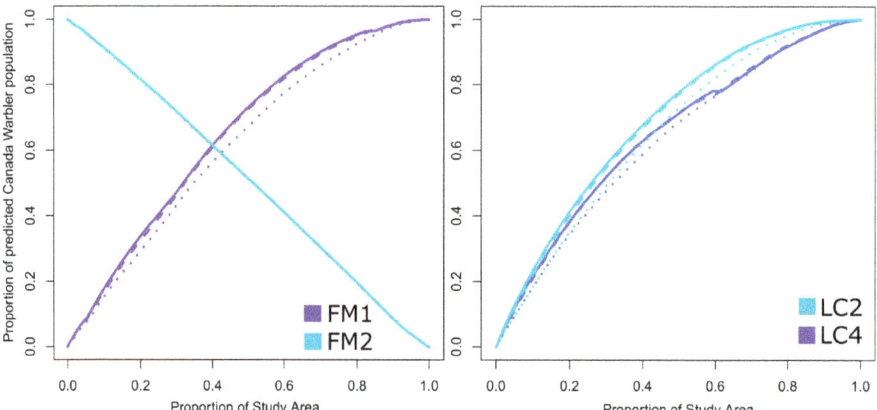

Figure 5. Response curves comparing efficiency (proportion of Canada Warbler population density retained per unit area) for two forest management scenarios (left panel) and two land conservation scenarios (right panel). Scenario descriptions given in Table 1. Curves shown at three dispersal distances: 5 km (LDD, solid line), 10 km (MDD, hashed line), and 50 km (HDD, dotted line).

For land conservation scenarios, cell-level rankings, which range from 0 to 1, differed more between MDD and HDD scenarios (mean = 0.10, SD = 0.90) than between LDD and MDD scenarios (mean = 0.04, SD = 0.04), with greater variation in differences also found between MDD and HDD (Table 3).

Table 3. Percent change in mean cell-level rankings across Zonation prioritization scenarios to support land conservation for the Canada Warbler when comparing between different estimated dispersal distances: 5 km (LDD), 10 km (MDDs), and 50 km (HDD). Scenario descriptions given in Table 1.

Scenario	Dispersal Estimates Compared	Mean Percent Change in Priority Ranking of a Given Cell between Scenarios	Standard Deviation of Percent Change in Priority Ranking
LC1	MDD-LDD	3	3
LC1	HDD-MDD	9	7
LC2	MDD-LDD	4	4
LC2	HDD-MDD	12	9
LC3	MDD-LDD	3	4
LC3	HDD-MDD	7	7
LC4	MDD-LDD	4	5
LC4	HDD-MDD	12	11

4. Discussion

We generated land conservation and forest management prioritization maps for the Canada Warbler in the Canadian portion of the Atlantic Northern Forest. This exercise identified several consistently prioritized areas for a range of stewardship objectives. In particular, central New Brunswick and southern Québec emerged as important areas for conservation and management under both current and future climate scenarios and when considering a range of possible dispersal distances. In general, areas farther from the coast were more frequently prioritized for conservation. Including projected effects of climate change on potential population density had dramatic effects in the scenarios, leading to almost no areas prioritized for conservation in Nova Scotia, and shifting priority areas northward. Within forest management tenures, the northern Gaspé Peninsula of Québec was consistently identified as the area where forest harvesting activities may be most practical while avoiding impacts to current Canada Warbler populations. This is consistent with Sólymos et al. [44], who predicted a lower average Canada Warbler population density in the Gaspé Peninsula compared to the rest of the study area.

Given this species' high conservation concern, our single-species exercise has potential to aid the rapid implementation of conservation and recovery action. Our method was somewhat unique in that landscape prioritization exercises are typically used for the assessment of biodiversity at large scales considering many different species (e.g., [19,51]). For a given taxon, Zonation results that are produced using only survey data could be different from those using species distribution models [52] and including both as input features may offer benefits. Although in our case adding recent locations of Canada Warbler only had a small impact on the overall solution, important differences were apparent at the level of individual cells (1 km²). Including this element is critical for land managers making decisions on small scales, as they indicate land parcels with persistent occupancy by Canada Warblers. These areas can thus be targeted for permanent land conservation and avoided during forest harvesting.

Although our prioritization scenarios provide insight into possible locations to target conservation and management activities, the spatial resolution of the scenarios (1 km) is too coarse to identify specific habitat patches. While two finer-scale Canada Warbler species distribution models have been constructed in this region [10,53], their coverage does not include the entire study area. Prioritized areas should be regarded as suggestions that require more detailed investigation through comparison with satellite imagery and ground-truthing before being considered candidates for conservation or management activity. To maximize efficacy, our analysis should be repeated as finer-scale spatial data become available to support local management planning and to account for the small territory sizes of the Canada Warbler (average 1 ha; [16,54]).

4.1. Accounting for Uncertainty

The predictions made by species distribution models, frequently used as input features in conservation prioritization exercises, are inherently more uncertain in under-sampled areas [55,56]. We accounted for this uncertainty by discounting densities by the standard deviation of mean population density, and thus were able to focus priorities on areas of lower scenario uncertainty [24,57].

We also attempted to account for uncertainty about dispersal distances within Canada Warbler populations by evaluating each scenario using three different dispersal distance estimates. These estimates were intended to influence results by dictating the extent to which priority conservation and management areas were required to be in close proximity to protected areas and known Canada Warbler locations, and to influence the allowed distance between current and future projected Canada Warbler distributions. However, we also used dispersal distance estimates within Zonation's 'Distribution Smoothing' function, which aggregates priority areas based on the assumption that fragmented solutions are undesirable [18], thereby yielding results that were increasingly spatially aggregated with larger dispersal distance estimates. This led to large differences in geographic priorities across scenarios, suggesting the need for careful consideration of habitat connectivity requirements for Canada Warbler and other species of conservation concern. Given that dispersal is a key parameter in spatial conservation prioritization [50], our findings highlight the importance of considering a range of dispersal assumptions. Future studies may benefit from multiple scenarios designed to isolate the different ways in which dispersal estimates can influence results.

The increased level of spatial aggregation in conservation priorities with higher dispersal estimates led to less efficient solutions in terms of the current predicted Canada Warbler population that would be conserved per unit area. However, conservation efficiency was influenced less by assumptions about dispersal distance than by assumptions about climate change effects on species' distributions, due to the large discrepancies between predicted current and future suitable habitats for boreal species [17]. Our results are consistent with those of Stralberg et al. [24], who found that incorporating potential avian responses to climate change reduced conservation efficiency for current songbird populations. This supports the idea that planning to incorporate climate change increases the size required for protected areas to adequately conserve species [17,58,59].

4.2. Maintaining Viable Populations on the Landscape

Connectivity was important in identifying management opportunities through this prioritization effort because Canada Warblers cluster in multi-territory "neighborhoods" [16,60]. Canada Warblers use forest stands post-harvest more often if they are within 100 m of unharvested stands with conspecific breeders; in one study, the presence of conspecifics was found to be more important than habitat condition for predicting stand use [16]. Therefore, it is important to conserve or manage for areas of sufficient size to support a neighborhood, although the ideal size and configuration of such habitat patches is not yet known. It is also not known whether dispersal within and between such neighborhoods is important for Canada Warbler population viability, nor what barriers to functional connectivity [61] exist for this species in this region. However, observed population declines combined with stable breeding productivity [62] suggest that eastern populations may not be limited by quality or connectivity of breeding habitat.

Our scenarios that assumed a high dispersal distance (HDD) prioritized large areas in the northwest portion of the study region. In contrast, the low dispersal distance (LDD) scenarios prioritized more locations of smaller size across the Atlantic Northern Forest. Relying on the results from HDD scenarios alone could undermine the goals of individual provinces to maintain native species by favoring the conservation of fewer large populations rather than a larger number of small, spatially distributed populations. The latter may be preferable from the standpoint of maintaining genetic variability across the landscape [63]. Furthermore, our scenarios that considered future climate projections (particularly the HDD variants) assigned low priority to southern populations in Nova Scotia and southern Québec—areas that become less hospitable to Canada Warbler in a warmer climate. Thus, the LDD and current climate scenarios represent more conservative assumptions for conservation and management. This may be more appropriate for a species that is experiencing population declines, especially given range-wide projected increases in habitat suitability under climate change [17].

To guarantee long-term persistence of high-quality breeding areas for the Canada Warbler, information on the capacity of habitats to support viable populations through detailed spatial population viability analyses is critical [64]. Although Zonation uses connectivity of populations as a surrogate for viability [20], this may not be as relevant for passerines, who have greater mobility than taxa with small dispersal distances such as small mammals, plants, or colonial birds. By incorporating measures of connectivity and including recent and repeated observations of Canada Warblers to locate high priority areas, we may have been able to better prioritize the landscape for conservation of high-value habitat than what was accomplished by using species distribution models or survey data alone. We were not able to include data on habitat selection and reproductive success that may have given a more direct indicator of population viability, which is being used in ongoing studies (Burns & Reitsma, in revision; Amelie Roberto-Charron, pers. comm.; Junior Tremblay, pers. comm.). Future prioritization efforts should include results of assessments of reproductive success and population viability wherever possible in order to most accurately target areas to maintain population density on the landscape. Such data would be particularly valuable if comparing harvested and unharvested areas, to test the hypothesis of whether areas undergoing forest management represent ecological traps for this species [65].

The present study advances the understanding of conservation issues and management opportunities for Canada Warbler at a regional scale. Our results help to define priority areas for Canada Warbler land conservation and forest management, and in conjunction with the habitat guidelines for the Canadian portion of the Atlantic Northern Forest [13] (see Supplementary Materials), they provide a toolkit for managers to immediately locate areas for implementing conservation and management actions. We suggest that this approach, designed to support management objectives for a single species, be applied to other species. We particularly encourage managers to apply this prioritization approach to Canada Warbler populations in other BCRs in Canada and the U.S. to support persistence of the entire breeding population.

Supplementary Materials: The following are available online at http://www.mdpi.com/1424-2818/12/2/61/s1, file 1: Westwood et al—2017—Guidelines for managing Canada Warbler habitat.pdf; file 2: Westwood, Reitsma, Lambert—2017—Prioritizing areas for Canada Warbler conservation and management.pdf.

Author Contributions: Conceptualization, J.D.L., A.R.W., and L.R.R.; methodology—development, spatial analysis, and statistics, A.R.W. and D.S.; writing—original draft preparation, review and editing, A.R.W., J.D.L., L.R.R., and D.S.; funding acquisition, J.D.L. and A.R.W. All authors have read and agree to the published version of the manuscript.

Funding: This research was funded by Environment and Climate Change Canada (ECCC), the U.S. Fish and Wildlife Service (USFWS), and High Branch Conservation Services Ltd. Article Processing Charges were funded by [TBA].

Acknowledgments: We acknowledge our funders for their support through various components of this project. Thank you to Alaine Camfield and Sean LeMoine of ECCC who were highly involved in the preparation of associated technical reports. The associated technical reports acknowledge the names of the many professionals across sectors (government, industry, academia, and Indigenous communities) who responded to consultations and surveys associated with this project, and for whose time and expertise we are grateful. We acknowledge the BAM Project's members, avian and biophysical data partners, and funding agencies, listed in full at www.borealbirds.ca/index.php/acknowledgements. The BAM project's database includes Breeding Bird Survey data and provincial Breeding Bird Atlas data, and we acknowledge the hundreds of skilled volunteers in Canada who have participated in these surveys over the years, and those who have served as provincial or territorial coordinators for the BBS. We thank the subject editor and two reviewers for a previous submission to the journal Avian Conservation and Ecology for their constructive comments, which refined the manuscript and led us to find a more appropriate home for applied conservation work. We are grateful to BAM members Péter Sólymos, and Trish Fontaine for provision of data and Nicole Barker and Samuel Haché for guidance and feedback. Finally, we thank Kara Pearson for GIS database assembly and Charlotte Harding for contributions to the U.S. habitat management guidelines.

Conflicts of Interest: The authors declare no conflict of interest.

References

1. Reitsma, L.R.; Goodnow, M.; Hallworth, M.T.; Conway, C.J. Canada Warbler (*Cardellina canadensis*). In *The Birds of North America Online*; Poole, A., Ed.; Cornell Lab of Ornithology: Ithaca, NY, USA, 2010.
2. Sauer, J.R.; Hines, J.E.; Pardieck, K.L.; Ziolkowski, D.J., Jr.; Link, W.A. *The North American Breeding Bird Survey, Results and Analysis 1966–2013*; Version 02.19; USGS Patuxent Wildlife Research Center: Laurel, MD, USA, 2015.
3. Government of Canada Species At Risk Act. S.C. 2002, c.29. Available online: http://www.sararegistry.gc.ca/default_e.cfm (accessed on 15 October 2018).
4. Maine Dept. of Inland Fisheries and Wildlife. *Maine's Wildlife Action Plan*; Maine Dept. of Inland Fisheries and Wildlife: August, ME, USA, 2015.
5. Kennedy, J.; Cheskey, T. Notes from the NAOC VI CWICI Roundtable. In Proceedings of the Canada Warbler International Conservation Initiative Roundtable, Washington, DC, USA, 18 August 2016; pp. 1–6.
6. Environment and Climate Change Canada North American Breeding Bird Survey—Canadian Trends Website, Data-Version. 2017. Available online: https://wildlife-species.canada.ca/breeding-bird-survey-results/P004/A001/?lang=e&m=s&r=OSFL&p=S (accessed on 2 February 2020).
7. Westwood, A.R. Conservation of three forest landbird species at risk: Characterizing and modelling habitat at multiple scales to guide management Planning. Ph.D. Thesis, Department of Biology, Dalhousie University, Halifax, NS, Canada, 2016.
8. Goodnow, M.; Reitsma, L.R. Nest-site selection in the Canada Warbler (*Wilsonia canadensis*) in central New Hampshire. *Can. J. Zool.* **2011**, *89*, 1172–1177. [CrossRef]
9. Hallworth, M.; Ueland, A.; Anderson, E.; Lambert, J.D.; Reitsma, L.R. Habitat selection and site fidelity of Canada Warblers (*Wilsonia canadensis*) in Central New Hampshire. *Auk* **2008**, *125*, 880–888. [CrossRef]
10. Westwood, A.R.; Staicer, C.; Sólymos, P.; Haché, S.; Fontaine, T.; Bayne, E.M.; Mazerolle, D.; Solymos, P.; Hache, S.; Fontaine, T.; et al. Estimating the conservation value of protected areas in Maritime Canada for two species at risk: the Olive-sided Flycatcher (*Contopus cooperi*) and Canada Warbler (*Cardellina canadensis*). *Avian Conserv. Ecol.* **2019**, *14*, 16. [CrossRef]
11. Grinde, A.R.; Niemi, G.J. Influence of landscape, habitat, and species co-occurrence on occupancy dynamics of Canada Warblers. *Condor* **2016**, *118*, 513–531. [CrossRef]
12. Hallworth, M.; Benham, P.M.; Lambert, J.D.; Reitsma, L.R. Canada warbler (*Wilsonia canadensis*) breeding ecology in young forest stands compared to a red maple (*Acer rubrum*) swamp. *For. Ecol. Manag.* **2008**, *255*, 1353–1358. [CrossRef]
13. Westwood, A.R.; Harding, C.; Reitsma, L.; Lambert, D. *Guidelines for Managing Canada Warbler Habitat in the Atlantic Northern Forest of Canada*; High Branch Conservation Services: Hartland, VT, USA, 2017.
14. Askins, R.A.; Philbrick, M.J. Effect of changes in regional forest abundance on the decline and recovery a forest bird community. *Wilson Bull.* **1987**, *99*, 7–21.
15. Bayne, E.; Leston, L.; Mahon, C.L.; Sólymos, P.; Machtans, C.; Lankau, H.; Ball, J.R.; Van Wilgenburg, S.L.; Cumming, S.G.; Fontaine, T.; et al. Boreal bird abundance estimates within different energy sector disturbances vary with point count radius. *Condor* **2016**, *118*, 376–390. [CrossRef]
16. Hunt, A.R.; Bayne, E.M.; Haché, S. Forestry and conspecifics influence Canada Warbler (*Cardellina canadensis*) habitat use and reproductive activity in boreal Alberta, Canada. *Condor* **2017**, *119*, 832–847. [CrossRef]
17. Stralberg, D.; Bayne, E.M.; Cumming, S.G.; Sólymos, P.; Song, S.J.; Schmiegelow, F.K.A. Conservation of future boreal forest bird communities considering lags in vegetation response to climate change: A modified refugia approach. *Divers. Distrib.* **2015**, *21*, 1112–1128. [CrossRef]
18. Moilanen, A.J.; Pouzols, F.M.; Meller, L.; Veach, V.; Arponen, A.; Leppänen, J.; Kujala, H. *Spatial conservation planning methods and software: ZONATION 4.0 User Manual*; Version 4; Conservation Biology Informatics Group, University of Helsinki: Helsinki, Finland, 2014; ISBN 9789521099199.

19. Ball, I.; Possingham, H.P.; Watts, M. Marxan and relatives: Software for spatial conservation prioritisation. In *Spatial Conservation Prioritisation: Quantitative Methods and Computational Tools*; Moilanen, A., Wilson, K., Possingham, H., Eds.; Oxford University Press: Oxford, UK, 2009; pp. 185–195.

20. Early, R.; Thomas, C.D. Multispecies conservation planning: Identifying landscapes for the conservation of viable populations using local and continental species priorities. *J. Appl. Ecol.* **2007**, *44*, 253–262. [CrossRef]

21. Shriner, S.A.; Wilson, K.R.; Flather, C.H. Reserve networks based on richness hotspots and representations vary with scale. *Ecol. Appl.* **2006**, *16*, 1660–1673. [CrossRef]

22. Franco, A.M.A.; Anderson, B.J.; Roy, D.B.; Gillings, S.; Fox, R.; Moilanen, A.J.; Thomas, C.D. Surrogacy and persistence in reserve selection: Landscape prioritization for multiple taxa in Britain. *J. Appl. Ecol.* **2009**, *46*, 82–91. [CrossRef]

23. Mikkonen, N.; Moilanen, A.J. Identification of top priority areas and management landscapes from a national Natura 2000 network. *Environ. Sci. Policy* **2013**, *27*, 11–20. [CrossRef]

24. Stralberg, D.; Camfield, A.; Carlson, M.; Lauzon, C.; Westwood, A.R.; Barker, N.K.S.; Song, S.J.; Schmiegelow, F.K.A. Strategies for identifying priority areas for songbird conservation in Canada's boreal forest. *Avian Conserv. Ecol.* **2018**, *13*, 12. [CrossRef]

25. Moilanen, A.J.; Anderson, B.J.; Eigenbrod, F.; Heinemeyer, A.; Roy, D.B.; Gillings, S.; Armsworth, P.R.; Gaston, K.J.; Thomas, C.D. Balancing alternative land uses in conservation prioritization. *Ecol. Appl.* **2011**, *21*, 1419–1426. [CrossRef]

26. Kareksela, S.; Moilanen, A.J.; Tuominen, S.; Kotiaho, J.S. Use of inverse spatial conservation prioritization to avoid biological diversity loss outside protected areas. *Conserv. Biol.* **2013**, *27*, 1294–1303. [CrossRef]

27. Kukkala, A.S.; Moilanen, A.J. Ecosystem services and connectivity in spatial conservation prioritization. *Landsc. Ecol.* **2017**, *32*, 5–14. [CrossRef]

28. Paradis, E.; Baillie, S.R.; Sutherland, W.J.; Gregory, R.D. Patterns of natal and breeding dispersal in birds. *J. Anim. Ecol.* **1998**, *67*, 518–536. [CrossRef]

29. Sutherland, G.D.; Harestad, A.S.; Price, K.; Lertzman, K.P. Scaling of natal dispersal distances in terrestrial birds and mammals. *Ecol. Soc.* **2000**, *4*, 1. [CrossRef]

30. Tittler, R.; Villard, M.A.; Fahrig, L. How far do songbirds disperse? *Ecography (Cop.).* **2009**, *32*, 1051–1061. [CrossRef]

31. Garrard, G.E.; McCarthy, M.A.; Vesk, P.A.; Radford, J.Q.; Bennett, A.F. A predictive model of avian natal dispersal distance provides prior information for investigating response to landscape change. *J. Anim. Ecol.* **2012**, *81*, 14–23. [CrossRef] [PubMed]

32. Game, E.T.; Kareiva, P.; Possingham, H.P. Six common mistakes in conservation priority setting. *Conserv. Biol.* **2013**, *27*, 480–485. [CrossRef] [PubMed]

33. Westwood, A.; Reitsma, L.R.; Lambert, D. *Prioritizing Areas for Canada Warbler Conservation and Management in the Atlantic Northern Forest of Canada*; High Branch Conservation Services: Hartland, VT, USA, 2017.

34. Moilanen, A.J. Landscape Zonation, benefit functions and target-based planning: Unifying reserve selection strategies. *Biol. Conserv.* **2007**, *134*, 571–579. [CrossRef]

35. Rich, T.D.; Beardmore, C.; Berlanga, H.; Blancher, P.J.; Bradstreet, M.; Butcher, G.; Demerast, D.; Dunn, E.; Hunter, W.; Iñigo-Elias, E.; et al. *Partners in Flight North American Landbird Conversation Plan*; Cornell Lab of Ornithology: Ithica, NY, USA, 2004.

36. Mahon, C.L.; Bayne, E.M.; Sólymos, P.; Matsuoka, S.M.; Carlson, M.; Dzus, E.; Schmiegelow, F.K.A.; Song, S.J. Does expected future landscape condition support proposed population objectives for boreal birds? *For. Ecol. Manag.* **2014**, *312*, 28–39. [CrossRef]

37. Ball, J.R.; Sólymos, P.; Schmiegelow, F.K.A.; Hache, S.; Schieck, J.; Bayne, E. Regional habitat needs of a nationally listed species, Canada Warbler (*Cardellina canadensis*), in Alberta, Canada. *Avian Conserv. Ecol.* **2016**, *11*, 10. [CrossRef]

38. Environment and Climate Change Canada Bird Conservation Strategies for Bird Conservation Region 14. Available online: https://www.canada.ca/en/environment-climate-change/services/migratory-bird-conservation/regions-strategies/description-region-14.html (accessed on 19 February 2018).

39. Trombulak, S.; Anderson, M.G.; Baldwin, R.; Beazley, K.; Ray, J.; Reining, C.; Woolmer, G.; Bettigole, C.; Forbes, G.; Gratton, L. *The Northern Appalachian/Acadian Ecoregion Priority Locations for Conservation Action*; Two Countries, One Forest Special Report; Two Countries, One Forest: Warner, NH, USA, 2008; No. 1.

40. Vankerham, R. Conservation Areas Reporting and Tracking System Database. Available online: http://www.ccea.org/carts/ (accessed on 2 February 2020).

41. Barker, N.K.S.; Fontaine, P.C.; Cumming, S.G.; Stralberg, D.; Westwood, A.R.; Bayne, E.M.; Sólymos, P.; Schmiegelow, F.K.A.; Song, S.J.; Rugg, D.J. Ecological monitoring through harmonizing existing data: Lessons from the boreal avian modelling project. *Wildl. Soc. Bull.* **2015**, *39*, 480–487. [CrossRef]

42. Sólymos, P.; Matsuoka, S.M.; Bayne, E.M.; Lele, S.R.; Fontaine, P.; Cumming, S.G.; Stralberg, D.; Schmiegelow, F.K.A.; Song, S.J. Calibrating indices of avian density from non-standardized survey data: making the most of a messy situation. *Methods Ecol. Evol.* **2013**, *4*, 1047–1058. [CrossRef]

43. Matsuoka, S.M.; Bayne, E.M.; Sólymos, P.; Fontaine, P.C.; Cumming, S.G.; Schmiegelow, F.K.A.; Song, S.J. Using binomial distance-sampling models to estimate the effective detection radius of point-count surveys across boreal Canada. *Auk* **2012**, *129*, 268–282. [CrossRef]

44. Haché, S.; Solymos, P.; Bayne, E.M.; Fontaine, T.; Cumming, S.G.; Schmiegelow, F.; Stralberg, D. Analyses to Support Critical Habitat Identification for Canada Warbler, Olive-sided Flycatcher, and Common Nighthawk: Final Report 1. Available online: https://zenodo.org/record/2433885#.XjZUmiMRVPY (accessed on 14 December 2019).

45. Stralberg, D.; Matsuoka, S.M.; Hamann, A.; Bayne, E.M.; Solymos, P.; Schmiegelow, F.K.A.; Wang, X.; Cumming, S.G.; Song, S.J. Projecting boreal bird responses to climate change: the signal exceeds the noise. *Ecol. Appl.* **2015**, *25*, 52–69. [CrossRef]

46. Meehl, G.A.; Covey, C.; Delworth, T.; Latif, M.; McAvaney, B.; Mitchell, J.F.B.; Stouffer, R.J.; Taylor, K.E. The WCRP CMIP3 multimodel dataset: A new era in climatic change research. *Bull. Am. Meteorol. Soc.* **2007**, *88*, 1383–1394. [CrossRef]

47. Global Forest Watch Canada's Industrial Concessions. 2016. Available online: http://www.globalforestwatch.ca/node/273 (accessed on 2 February 2020).

48. Anderson, M.G.; Clark, M.; Ferree, C.; Jospe, A.; Sheldon Olivero, A.; Weaver, K. *Northeastern Habitat Guides: A Companion to the Terrestrial and Aquatic Habitat Maps*; The Nature Conservancy, Eastern Conservation Science, Eastern Regional Office: Boston, MA, USA, 2013.

49. Esri Inc. ArcGIS for Desktop 10.2.2. Available online: http://www.esri.com (accessed on 2 February 2020).

50. Carroll, C.; Dunk, J.R.; Moilanen, A.J. Optimizing resiliency of reserve networks to climate change: Multispecies conservation planning in the Pacific Northwest, USA. *Glob. Chang. Biol.* **2010**, *16*, 891–904. [CrossRef]

51. Betts, M.G.; Zitske, B.P.; Hadley, A.S.; Diamond, A.W. Migrant forest songbirds undertake breeding dispersal following timber harvest. *Northeast. Nat.* **2006**, *13*, 531–536. [CrossRef]

52. Rushing, C.S.; Ryder, T.B.; Marra, P.P.; Rushing, C.S. Quantifying drivers of population dynamics for a migratory bird throughout the annual cycle. *Proc. R. Soc. B Biol. Sci.* **2016**, *283*, 20152846. [CrossRef] [PubMed]

53. Bale, S.; Beazley, K.; Westwood, A.R.; Bush, P. The importance of using topographic features to predict climate-resilient habitat for migratory forest landbirds: An example for the Rusty Blackbird, Olive-sided Flycatcher, and Canada Warbler. *Condor* **2020**, *122*, 1–19.

54. Reitsma, L.R.; Hallworth, M.T.; Benham, P.M. Does age influence territory size, habitat selection, and reproductive success of male Canada Warblers in central New Hampshire? *Wilson J. Ornithol.* **2008**, *120*, 446–454. [CrossRef]

55. Araújo, M.B.; Peterson, A.T. Uses and misuses of bioclimatic envelope modeling. *Ecology* **2012**, *93*, 1527–1539. [CrossRef]

56. Moilanen, A.; Runge, M.C.; Elith, J.; Tyre, A.; Carmel, Y.; Fegraus, E.; Wintle, B.A.; Burgman, M.; Ben-Haim, Y. Planning for robust reserve networks using uncertainty analysis. *Ecol. Modell.* **2006**, *199*, 115–124. [CrossRef]

57. Moilanen, A.J.; Wintle, B.A. The boundary-quality penalty: A quantitative method for approximating species responses to fragmentation in reserve selection. *Conserv. Biol.* **2007**, *21*, 355–364. [CrossRef]

58. Hilty, J.A.; Chester, C.C.; Cross, M.S. (Eds.) *Climate and Conservation: Landscape and Seascape Science, Planning, and Action*; Island Press: Washington, DC, USA, 2012; ISBN 9781610911702.

59. Andrew, M.E.; Wulder, M.; Cardille, J.A. Protected areas in boreal Canada: A baseline and considerations for the continued development of a representative and effective reserve network. *Environ. Rev.* **2014**, *22*, 1–26. [CrossRef]

60. Reitsma, L.R.; Jukosky, J.A.; Kimiatek, A.J.; Goodnow, M.L.; Hallworth, M.T. Extra-pair paternity in a long-distance migratory songbird beyond neighbors' borders and across male age classes. *Can. J. Zool.* **2018**, *96*, 49–54. [CrossRef]

61. Robertson, O.J.; Radford, J.Q. Gap-crossing decisions of forest birds in a fragmented landscape. *Austral Ecol.* **2009**, *34*, 435–446. [CrossRef]

62. Wilson, S.; Saracco, J.F.; Krikun, R.G.; Flockhart, D.T.T.; Godwin, C.M.; Foster, K.R. Drivers of demographic decline across the annual cycle of a threatened migratory bird. *Sci. Rep.* **2018**, *8*, 7316. [CrossRef] [PubMed]

63. Akçakaya, H.R. Linking population-level risk assessment with landscape and habitat models. *Sci. Total Environ.* **2001**, *274*, 283–291. [CrossRef]

64. McCarthy, M.A. Spatial population viability analysis. In *Spatial Conservation Prioritization*; Moilanen, A., Wilson, K.A., Possingham, H.P., Eds.; Oxford University Press: New York, NY, USA, 2009; pp. 123–134.

65. Pärt, T.; Arlt, D.; Villard, M.-A. Empirical evidence for ecological traps: a two-step model focusing on individual decisions. *J. Ornithol.* **2007**, *148*, 327–332. [CrossRef]

Article

Occupancy of the American Three-Toed Woodpecker in a Heavily-Managed Boreal Forest of Eastern Canada

Vincent Lamarre [1] and Junior A. Tremblay [1,2,*]

[1] Science and Technology Branch, Environment and Climate Change Canada, 1550 Avenue d'Estimauville, Québec, QC G1J 0C3, Canada; vincent.lamarre@canada.ca

[2] Département des Sciences du Bois et de la Forêt, Université Laval, 2405 rue de la Terrasse, Québec, QC G1V 0A6, Canada

* Correspondence: junior.tremblay@canada.ca

Abstract: The southern extent of the boreal forest in North America has experienced intensive human disturbance in recent decades. Among these, forest harvesting leads to the substantial loss of late-successional stands that include key habitat attributes for several avian species. The American Three-toed Woodpecker, *Picoides dorsalis*, is associated with continuous old spruce forests in the eastern part of its range. In this study, we assessed the influence of habitat characteristics at different scales on the occupancy of American Three-toed Woodpecker in a heavily-managed boreal landscape of northeastern Canada, and we inferred species occupancy at the regional scale. We conducted 185 playback stations over two breeding seasons and modelled the occupancy of the species while taking into account the probability of detection. American Three-toed Woodpecker occupancy was lower in stands with large areas recently clear-cut, and higher in landscapes with large extents of old-growth forest dominated by black spruce. At the regional scale, areas with high probability of occupancy were scarce and mostly within protected areas. Habitat requirements of the American Three-toed Woodpecker during the breeding season, coupled with overall low occupancy rate in our study area, challenge its long-term sustainability in such heavily managed landscapes. Additionally, the scarcity of areas of high probability of occupancy in the region suggests that the ecological role of old forest outside protected areas could be compromised.

Keywords: boreal forest; clear-cutting; conservation; forest management; old-growth forest; *Picoides dorsalis*; protected areas

Citation: Lamarre, V.; Tremblay, J.A. Occupancy of the American Three-Toed Woodpecker in a Heavily-Managed Boreal Forest of Eastern Canada. *Diversity* **2021**, *13*, 35. https://doi.org/10.3390/d13010035

Received: 17 December 2020
Accepted: 15 January 2021
Published: 19 January 2021

Publisher's Note: MDPI stays neutral with regard to jurisdictional claims in published maps and institutional affiliations.

1. Introduction

The boreal forest represents about 48% of the world's forested biomes [1]. Boreal landscapes are shaped by natural disturbances such as windthrows, forest fires and insect outbreaks that generate a complex mosaic in vegetation structure and composition [2,3]. For several decades, however, the exploitation of natural resources has been adding to natural disturbances in this biome [4] and has modified the structure and composition of boreal ecosystems, including the reduction of late-successional stands called old-growth forests [5,6].

Old-growth boreal forests are characterized, among several attributes, by irregular vertical and horizontal structures, large volumes of deadwood either standing (snags) or fallen (coarse woody debris) and in different decaying stages [7]. These attributes are considered key for hundreds of species that depend upon dead or decaying woody material during some part of their life cycle and that are found disproportionately in old-growth forests [8,9]. The temporal continuity of these ecological attributes is also an essential characteristic for many species [10,11]. For instance, a recent study highlights the continuous supply of large slightly decayed snags in specific old-growth forest type as a key element to provide temporal stability in the foraging habitat of the Black-backed Woodpecker, *Picoides arcticus* [12].

45

Deadwood provides foraging or breeding substrate for several vertebrate species [13]. Boreal woodpeckers, for example, are "ecosystem engineers" that create nesting cavities for other vertebrates and are considered indicator species for deadwood-associated biodiversity [14,15]. Among woodpecker species found within the boreal biome in North America, the American Three-toed Woodpecker, *Picoides dorsalis*, is the most strongly associated with continuous old spruce forests at the landscape scale [16–18]. Harvesting, especially of old-growth coniferous forests, is thought to be detrimental to the species by limiting the supply of deadwood [16,18–20]. However, most of the studies on the species in eastern Canada occurred in regions where forests were harvested for the first time and where mature and old stands were still relatively abundant amongst residual forests. In heavily-managed forest landscapes, the persistence of the species remains uncertain. Indeed, in a recent attempt to gain insight on the breeding ecology of American Three-toed Woodpecker in heavily managed forest at the southern edge of its breeding range (New-Brunswick, Canada), Craig et al. [21] reported the species in only 5.9% of the playback stations, and found no predictors of site occupancy although most nests were found in recently dead black spruce trees.

Here, we assess the influence of habitat characteristics at the stand and landscape scales on the occupancy of American Three-toed Woodpecker during the breeding period in a heavily-managed, unburned boreal forest landscape of eastern Canada, and infer the probability of occupancy of the species at the regional scale. We expect that American Three-toed Woodpecker occupancy would be favoured by old spruce forest stands but negatively affected by recently harvested forest stands and higher forest fragmentation, and that the probability of occupancy of the species at the regional scale would be low.

2. Materials and Methods

2.1. Study Area

This study was conducted at the Forêt Montmorency (Université Laval's Experimental Forest), and the Réserve Faunique des Laurentides, Québec, Canada (47°4′ N. 71°0′ W.; Figure 1). The study area covers approximately 210 km^2 within the balsam fir-white birch bioclimatic domain in the continuous boreal forest subzone [22], which represents the southernmost section of the boreal forest in eastern Canada [23]. More precisely, the study area is classified within the high-elevation balsam fir–white birch zone with elevation ranging between 600 and 1100 m [24]. Forest stands within the study area are dominated by balsam fir *Abies balsamea* and black spruce *Picea mariana*, the latter being more abundant at higher elevation. Companion species are white birch (*Betula papyrifera*), white spruce (*Picea glauca*), and tamarack (*Larix laricina*). In the region, average age-class structure of the natural variability observed over the last few centuries includes 86% of old forest stands [25]. However, intensive logging during the 20th and early 21st century led to a net reduction in the cover of mature and old forest stands [26]. During the 1909–2005 time frame, 105% of the high-elevation boreal landscape has been harvested with some areas logged twice [27]. Consequently, the studied area is now dominated by young regenerating stands (0–19 years; 39%), while the cover of old forests (>90 years) is approximately 25%. Young closed (20–59 years) and mature (60–89 years) forests as well as non-forested lands (e.g., lakes, wetlands, etc.) respectively cover 11, 16 and 9% of the study area. Dominant cover tree species of old forest stands are balsam fir (55%), black spruce (42%) and tamarack (3%).

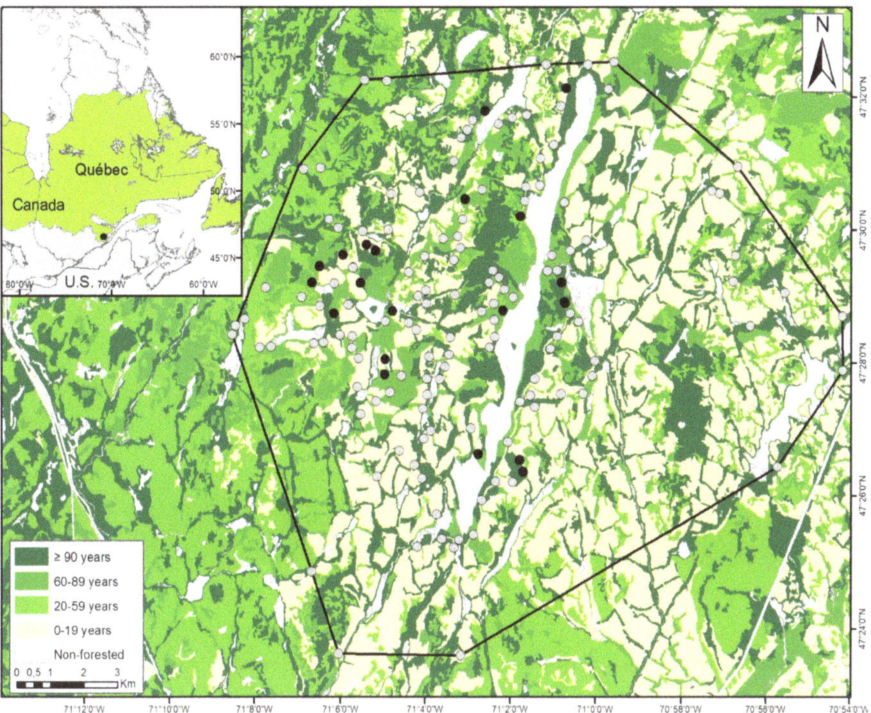

Figure 1. Habitat composition and location of surveyed stations where American Three-toed Woodpecker was detected (black circles) and not detected (grey circles) during occupancy surveys conducted between 2016 and 2017 within the Forêt Montmorency and the Réserve Faunique des Laurentides (Québec, Canada). Boundaries of the boreal zone [28] are delimited by the green zone in the left insert. The black line delineates the boundaries of the study area in the large insert.

2.2. Woodpecker Surveys

In 2016 and 2017, we conducted woodpecker occupancy surveys during American Three-toed Woodpecker breeding season at stationary stations distributed along forest roads. We did not stratify the sampling *per se* but rather targeted areas with clusters of unharvested patches within the vicinity of old forest stands. Surveyed stations were distanced by a minimum of 255 m (mean ± standard deviation: 554 ± 442 m; range 255–3414 m), while mean elevation at stations was 889 ± 37 m (range 819–1015 m). Surveys lasted five minutes (300 s), during which conspecific playbacks were displayed, and we noted the time elapsed since the start of the playback when an individual was observed or heard. Similar to the method used by Craig et al. [21], the 300-s conspecific display was divided into two 120-s observation periods which constituted the two visits used in occupancy analyses (see below). Both observation periods were interspersed by a 60-s pause. Playbacks were mixed from recordings obtained from xeno-canto [29] and the Macaulay Library [30]. Playbacks could be heard to approximately 450 m by the human ear in the field. We conducted surveys between 5:47 a.m. and 3:32 p.m., when precipitations were absent or minimal and wind speed $\leq$ 3 on the Beaufort scale (12–19 km h^{-1}). Afterward, we accessed three-hour mean wind speed from the nearest weather station located 18 km northwest of the study area for further analysis [31].

2.3. Habitat Characteristics

We investigated the influence of habitat characteristics that were likely to affect the occupancy of American Three-toed Woodpecker by calculating vegetation covariates within buffers of 250 and 750 m-radii centered on each surveyed station, representing respectively the stand and landscape scales. The landscape scale was selected based on home range size estimates obtained for one female and two male American Three-toed Woodpeckers during the nesting period in the study area, which averaged 177 ha (J.A. Tremblay, unpublished data). Vegetation covariates were calculated using the eco-forestry layers available as of 2017 for the study area [32]. At the stand scale, we calculated the area (ha) covered by recent clear-cut stands *clear_cut*. At the stand and landscape scales, we measured the area (ha) covered by old forest ($\geq$90 years) dominated by black spruce *old_spruce* and mean forest age *mean_age*. We weighted *mean_age* by the area covered by even-aged stands within the buffers around each surveyed station. We calculated the standard deviation of mean stand age *sd_age* as an index of habitat fragmentation at the landscape scale. Based on the literature in eastern Canada [16–18], the area covered by non-forested lands in the landscape, the area covered by young (20–59 years) and mature (60–89 years) forests at the stand and landscape scales have been considered marginal for the American Three-toed Woodpecker in the selection of its habitat during the breeding season and thus were excluded from occupancy analyses (see below). We nevertheless compared mean value of all habitat characteristics between stations where the American Three-toed Woodpecker was detected, not detected and a set of 185 random stations distributed randomly into the study (Table 1) area using a Kruskal–Wallis test in R statistical environment version 4.0.2 [33].

Table 1. Description and mean values $\pm$ standard deviation (s.d.) of habitat characteristics measured at the stand (250 m-radii) and landscape (750 m-radii) scales at stations where the American Three-toed Woodpecker was detected, not detected and at random stations during occupancy surveys in managed boreal forest in eastern Canada. Shared letters indicate no significant differences in habitat characteristics among stations (Kruskal–Wallis test). Habitat characteristics in bold are included in occupancy analyses.

Habitat Characteristic	Description	Not Detected (n = 163)	Detected (n = 22)	Random (n = 185)
clear_cut_250	Area (ha) covered by recent (0–19 years) clear-cut	6.58 (4.62) [a]	3.91 (4.17) [b]	6.60 (5.89) [a]
mean_age_250	Mean forest age (year)	70.78 (25.44) [a]	83.27 (30.27) [a]	64.92 (32.34) [b]
mean_age_750		70.07 (15.34) [a]	76.27 (20.63) [a]	64.33 (19.47) [b]
young_250	Area (ha) covered by young closed forest (20–59 years)	1.29 (2.96) [a]	0.86 (1.54) [a,b]	3.24 (5.06) [b]
young_750		14.71 (24.20) [a]	8.34 (10.62) [a]	27.03 (35.24) [b]
mature_250	Area (ha) covered by mature forest (60–89 years)	4.87 (4.64) [a]	6.23 (5.72) [a]	3.11 (3.80) [b]
mature_750		38.49 (21.80) [a]	48.89 (23.05) [a]	28.00 (22.32) [b]
old_spruce_250	Area (ha) covered by old forest ($\geq$90 years) dominated by black spruce	2.92 (3.04) [a]	3.85 (3.46) [a]	2.26 (3.51) [b]
old_spruce_750		21.87 (13.18) [a]	26.00 (16.33) [a]	19.92 (15.08) [a]
sd_age_750	Standard deviation of forest age	56.71 (8.89) [a]	53.08 (9.35) [a]	54.62 (9.80) [a]
nf_750	Area (ha) covered by non-forested lands in the landscape	21.04 (19.98) [a]	26.44 (23.93) [a]	12.00 (16.16) [b]

2.4. Occupancy Analyses

We used single-species occupancy modeling in the *unmarked* R library to investigate the influence of habitat characteristics on the probability of occupancy of American Three-toed Woodpecker [34,35]. We converted survey detections of American Three-toed Woodpecker into presence–absence data. To account for imperfect detection probability during surveys, we first estimated the effect of year, Julian day, wind speed (m s^{-1}) and hour of the day on detection probability of American Three-toed Woodpecker. We used a two-step approach [36] to determine which detection parameter(s) to retain in the occupancy models. We first estimated the effect of each detection parameter by ranking univariate models in which occupancy was held constant. We also included a null model in which detection and occupancy were held constant (Table A1). The five candidate models were ranked based on the second-order Akaike's information criterion AICc [37]; using the aictab function of the *AICcmodavg* R library [38].

We developed a set of 10 biologically relevant candidate occupancy models representing hypotheses to investigate the influence of habitat characteristics on the probability of occupancy of American Three-toed Woodpecker (Table 2). This set also included a null model of constant occupancy. Strongly correlated covariates ($|r| \geq 0.60$; Table A2) were not included in the same candidate model to reduce multicollinearity. We ranked candidate models based on the AIC_c and we made inference on the top models with $\Delta AICc < 2$. Parameter-averaged predictions of covariates appearing in the most parsimonious models were calculated over their measured range while holding other variables at their mean value. Finally, we inferred a predicted probability of occupancy of American Three-toed Woodpecker from the top models (Table 2) at the regional scale (30 km $\times$ 30 km square zone) within the balsam fir–white birch bioclimatic domain.

Table 2. Candidates models, number of parameters (k), second-order Akaike's information criterion (AICc), ΔAICc, Akaike weights (ω), log-likelihood (LL) of the candidate models assessing occupancy (ψ) and detection probability (p) of American Three-toed Woodpecker in a managed boreal forest in eastern Canada.

Candidate Model	k	AIC_c	ΔAIC_c	ω	LL
~ p(time) ~ ψ(clear_cut_250 + old_spruce_750)	5	176.56	0.00	0.49	−83.11
~ p(time) ~ ψ(clear_cut_250)	4	178.46	1.90	0.19	−85.12
~ p(time) ~ ψ(mean_age_250 + sd_age_750)	5	178.99	2.43	0.15	−84.33
~ p(time) ~ ψ(clear_cut_250 + mean_age_750)	5	180.22	3.65	0.08	−84.94
~ p(time) ~ ψ(mean_age_250)	4	182.46	5.89	0.03	−87.12
~ p(time) ~ ψ(old_spruce_250 + sd_age_750)	5	182.76	6.20	0.02	−86.21
~ p(time) ~ ψ(sd_age_750)	4	183.63	7.06	0.01	−87.70
~ p(time) ~ ψ(mean_age_250 + old_spruce_750)	5	183.86	7.30	0.01	−86.76
~ p(time) ~ ψ(null)	3	185.05	8.49	0.01	−89.46
~ p(time) ~ ψ(old_spruce_250)	4	185.25	8.68	0.01	−88.51

3. Results

Between 19 May and 13 July 2016 and between 26 May and 29 June 2017, we broadcasted American Three-toed Woodpecker playbacks at 185 stations, including 35 revisited stations in 2017. American Three-toed Woodpecker was detected at 22 (11.9%) of the 185 stations. Forest composition and structure differed between sampled and random stations since we targeted our sampling effort towards clusters of older residual forest patches distributed close to non-forested lands such as waterbodies and wetlands in the study area (Table 1). Indeed, mean forest age at the stand and landscape scale and the area covered by old spruce at the stand scale were lower at random than sampled stations. In addition, the area covered by mature stands at both scales was also lower at random stations, while the dominance of young stands tended to be greater. Finally, non-forested habitats in the landscape were less important at random than sampled stations (Table 1).

Time of the day was retained in the most parsimonious model and accounted for 42% of model selection weight (Table A1). The models including year ($\Delta AIC_c = 1.61$) and date ($\Delta AIC_c = 1.95$) were equivalent and did not differ from the null model ($\Delta AIC_c = 1.77$). Only time of the day influenced the probability of detection of American Three-toed Woodpecker (p_{time}: 0.37, 95% C.I.: [0.03, 0.71]) and therefore was the only detection parameter included in occupancy models. Mean occupancy rate of American Three-toed Woodpecker in the study site was 0.17 ± 0.04, and mean detection probability when the time of the day was fixed to its mean value was 0.48 ± 0.11.

Two models had a substantial level of empirical support in influencing occupancy of the American Three-toed Woodpecker (ΔAIC$_c$ < 2; Table 2). Together, these models accounted for 68% of the AIC weight. At the stand scale, occupancy of American Three-toed Woodpecker decreased with increasing recent clear-cut area ($\psi_{clear\_cut\_250}$: −0.20, 95% C.I.: [−0.34, −0.06], Figure 2a). The area covered by recent clear-cuts at the stand scale was also lower at stations where American Three-toed Woodpecker was detected compared to stations where the species was not detected or random stations (Table 1). At the landscape scale, occupancy of the species tended to increase with an increase in the area covered by old forest dominated by black spruce ($\psi_{old\_spruce\_750}$: 0.04, 95% C.I.: [0.00–0.08], Figure 2b), with increasing uncertainty in confidence intervals > 40 ha most likely due to the scarcity of large old spruce forest stands in our study area.

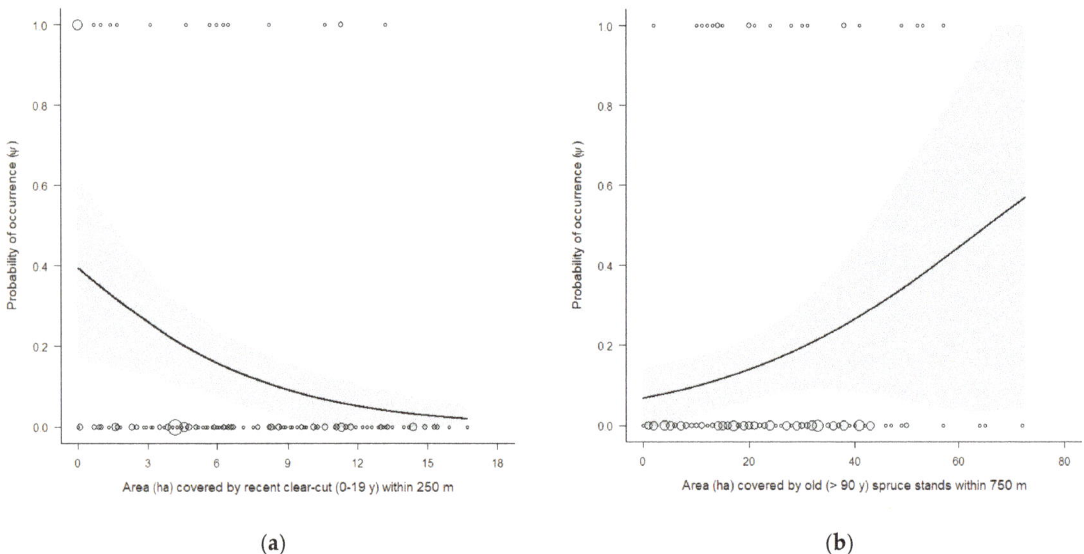

(a)   (b)

Figure 2. Influence on the occupancy of American Three-toed Woodpecker in managed boreal forest in eastern Canada of the area (ha) covered by (**a**) recent clear-cut stands at the stand scale (250 m radii) and (**b**) old spruce stands at the landscape scale (750 m radii). The shaded grey area represents 95% confidence interval, and symbols represent the distribution of raw data with symbol size proportional to the (log + 1) number of observations.

Extrapolating results inferred from our top occupancy models (Table 2) to a larger extent within the balsam fir–white birch bioclimatic domain showed that only 12.3% of the region had moderate- to high-predicted probability of occupancy (>0.5) of the American Three-toed Woodpecker. About half of these areas (6.7%) were located within protected areas. Areas with high-predicted probability of occupancy (>0.75) of the species represented 4.7% of the region where only 1.0% were in managed forests outside protected areas (Figure 3).

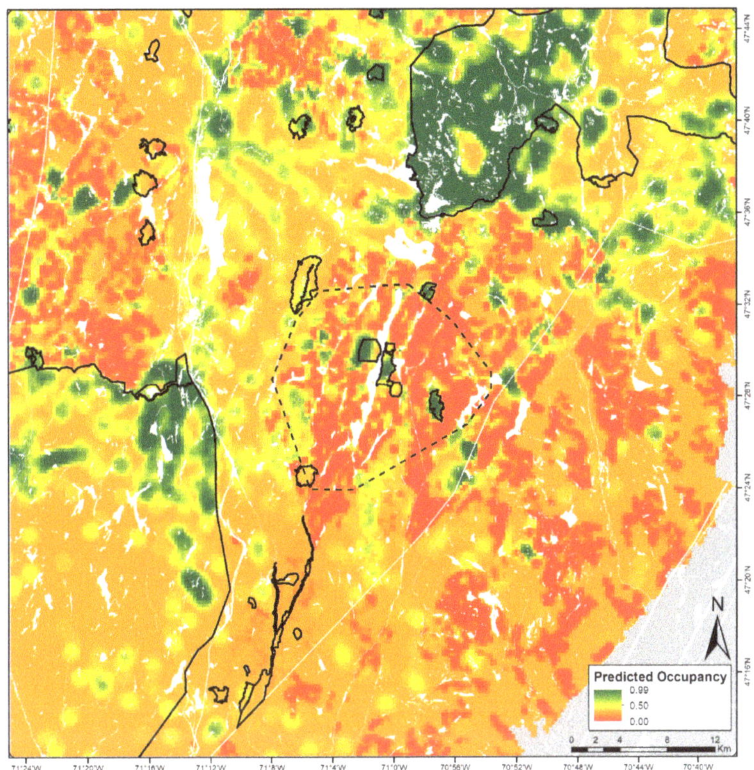

Figure 3. Predicted probability of occupancy of American Three-toed Woodpecker at the regional scale within the balsam fir–white birch bioclimatic domain. Probability of occupancy is inferred from the top-ranking models ($\Delta AIC_c < 2$; Table 2). The dashed and solid black lines respectively delineate the boundaries of the study area and protected areas and white polygons indicate non-forested lands.

4. Discussion

In a heavily-managed landscape at the southern edge of the boreal forest, the American Three-toed Woodpecker exhibited a low mean occupancy rate (0.17 ± 0.04) where its probability of occupancy decreased rapidly with increasing area of recent clear-cut at the stand scale. The amount of old spruce forest was positively associated with species occupancy at the landscape scale. Our results suggest that logging history in the region may have created an unsuitable forest landscape (i.e., dominated by younger age-classes and fragmented residual older forest stands) to support a long-term population of American Three-toed Woodpecker.

Habitat associations of the American Three-toed Woodpecker vary across its range. In a meta-analysis focusing on successional trajectories of bird communities following fire and harvesting in boreal forests of western Canada, Schiek and Song [39] report the American Three-toed Woodpecker more common in older mixed wood and white spruce. In western Quebec, Imbeau and Desrochers [16] document a positive association between the American Three-toed Woodpecker with the amount of old spruce forest in continuous coniferous forest. Similarly, Cadieux and Drapeau [18] report a higher occurrence of the species in black spruce forest stands older than 90 years and no occurrence in younger coniferous stands. Accordingly, we find that the occupancy of the American Three-toed Woodpecker during the breeding period increases with old spruce forest stands in a managed landscape. This result agrees with our prediction as it has been shown that the

American Three-toed Woodpecker preferentially forages on relatively large and senescent or recently dead coniferous trees to access bark-associated beetles (Scolytinae) by scaling layers of bark with strong preference for black spruce in eastern Canada [17,40,41].

Recent clear-cut at the stand scale have a negative effect on the occupancy of the American Three-toed Woodpecker. This result agrees with our expectation and with previous studies about the sensitivity of the species to forest harvesting, especially in the eastern part of its distribution range [19,20,41,42]. Accordingly, in the black spruce moss domain of eastern Canada, recent clear-cuts have a lower density of snags, and the American Three-toed Woodpecker is solely found in old-growth forest [19]. In addition to a reduction in the deadwood abundance, long-term recruitment of large-diameter snags is compromised in clear-cut stands [43]. In a closely-related species, the Black-backed Woodpecker, old coniferous forests with a greater volume of recently dead trees than adjacent recent cuts are selected for foraging during the breeding period [44]. Similar foraging avoidance of recent clear-cut areas may be occurring in our study site for the American Three-toed Woodpecker. Indeed, timber harvesting peaked during the period 1996–2004 and ended in 2009, and although we did not quantify the volume and decay stages of deadwood, it is reasonable to think that most deadwood in clear-cuts had entered late decay classes and was of low quality for foraging American Three-toedWoodpecker at the time of our study.

Our results do not report a predictive effect of habitat fragmentation at the landscape scale on the occupancy of the American Three-toed Woodpecker. Forest edge avoidance by foraging American Three-toed Woodpecker has been reported, where high-quality substrates near stand edges are used less frequently than available [41]. In addition, movements of foraging woodpeckers also appear to be constrained in residual forests following harvesting [16,41]. Hence, the effect of habitat fragmentation seems to act on a finer scale, and our results suggest that at a larger scale (i.e., landscape), the amount of old forests may strongly influence the occurrence of the species. For instance, most of our detection of the species occurred close to forest patches of mature or old forests rather than residual strips of forest (Figure 1).

Only 1% of the forest stands in managed forests at the regional scale show high predicted probability of occupancy of the American Three-toed Woodpecker. This is likely a consequence of forestry practices, mainly driven by clear-cutting during the last century which has led to a substantial reduction of old forest cover in the region [26,27]. Such practices can hardly sustain biodiversity associated with old coniferous forest stands, especially for species with large home ranges. For example, habitat alteration from anthropogenic activities is a threat of high concern for populations of Woodland Caribou *Rangifer tarandus caribou* across Canada's boreal biomes, including the local population in our study area, which is considered "not self-sustaining" [45]. Fennoscandia previously experienced a similar situation where old-growth forests have almost completely disappeared [46,47], and it is estimated that 30–50% of the red-listed species in these regions are associated with old-growth forest attributes [48,49]. In boreal ecosystems, conservation strategies cannot be based solely on a network of protected areas but rather on how unprotected areas are managed [50]. Hence, management practices mimicking the attributes of old-growth forests by ensuring a continuous recruitment of deadwood appear important for maintaining species associated with old spruce forests. Within these practices, partial harvesting may be efficient in maintaining a relatively high abundance of deadwood and associated deadwood-dependent species such as the American Three-toed Woodpecker [51,52], although the efficiency of such practices has not yet been assessed. The ecological role of old forests in our study area seems to be altered, and the situation may require revised management practices and a passive restoration of the ecological integrity of old forests outside protected areas in the region *sensu* [53].

5. Conclusions

With 185 stations sampled over two breeding seasons, our study on the occupancy of the American Three-toed Woodpecker is one of the first conducted in a heavily harvested landscape at the southern edge of the boreal forest. The overall low occupancy of the species within our study area raises questions about the long-term sustainability of such heavily managed landscapes for the American Three-toed Woodpecker. Areas of high-predicted occupancy of the species in the studied region are mostly found in protected areas, providing evidence that heavy forest harvesting is a detrimental driver that likely contributed to the significant long-term declining trends of the species over the 1970–2019 period in the province of Québec ($-0.96\%.\text{year}^{-1}$; 95% C.I.: $[-0.70\text{–}-1.22]$; [54]) and at a larger scale in the boreal hardwood transition bird conservation region ($-3.5\%.\text{year}^{-1}$; 95% C.I.: $[-2.0\text{–}-5.2]$; [55]). The southern part of the boreal forest in eastern Canada, where we conducted our study, has experienced one of the most important human disturbances in the past decades, mainly related to forest harvesting [4]. We pledge for more detailed studies on this discrete keystone species in different regions with varying forest harvesting intensity, at the stand scale (going from clear-cutting to partial harvesting) and landscape scale (from pristine to heavily harvested), focusing on demographic parameters (i.e., reproductive success and survival) and habitat selection.

Author Contributions: Conceptualization, J.A.T.; methodology, J.A.T.; formal analysis, J.A.T., V.L.; investigation, J.A.T., V.L.; resources, J.A.T.; data curation, V.L.; writing—original draft preparation, V.L.; writing—review and editing, J.A.T., V.L.; visualization, V.L.; supervision, J.A.T.; project administration, J.A.T.; funding acquisition, J.A.T. All authors have read and agreed to the published version of the manuscript.

Funding: This research was funded by Environment and Climate Change Canada.

Institutional Review Board Statement: Not applicable.

Informed Consent Statement: Not applicable.

Data Availability Statement: The data presented in this study are openly available in https://doi.org/10.5061/dryad.9kd51c5gc.

Acknowledgments: We are grateful to André Desrochers for sharing the code to calculate habitat characteristics and to Mark J. Mazerolle for advice related to statistical analyses. We are also thankful to the personnel at the Forêt Montmorency for their support and to Michel Robert for the loan of all-terrain vehicles. Finally, we thank Francis Lessard and Julien St-Amand for their help with fieldwork.

Conflicts of Interest: The authors declare no conflict of interest. The funders had no role in the design of the study; in the collection, analyses, or interpretation of data; in the writing of the manuscript, or in the decision to publish the results.

Appendix A

Table A1. Model selection based on the AIC_c for the estimation of detection probability (p) of American Three-toed Woodpecker in managed boreal forest in eastern Canada. The number of parameters (k), second-order Akaike's information criterion (AIC_c), ΔAIC_c, Akaike weights (ω), log-likelihood (LL) are included in the table and occupancy (ψ) is held constant for each candidate model.

Candidate Model	k	AIC_c	ΔAIC_c	ω	LL
~ p(time)	3	185.05	0.00	0.42	-89.46
~ p(year)	3	186.66	1.61	0.19	-90.27
~ p(null)	2	186.82	1.77	0.17	-91.38
~ p(date)	3	187.00	1.95	0.16	-90.44
~ p(wind)	3	188.60	3.55	0.07	-91.23

Table A2. Correlation among habitat characteristics included in occupancy analyses.

	mean_age_250	clear_cut_250	old_spruce_250	mean_age_750	sd_age_750	old_spruce_750
mean_age_250	1					
clear_cut_250	−0.53	1				
old_spruce_250	0.42	0.03	1			
mean_age_750	0.66	−0.36	0.19	1		
sd_age_750	0.19	0.31	0.32	0.28	1	
old_spruce_750	0.29	0.17	0.60	0.37	0.23	1

References

1. Brandt, J.P.; Flannigan, M.D.; Maynard, D.G.; Thompson, I.D.; Volney, W.J.A. An Introduction to Canada's Boreal Zone: Ecosystem Processes, Health, Sustainability, and Environmental Issues. *Environ. Rev.* **2013**, *21*, 207–226. [CrossRef]
2. Kneeshaw, D.D.; Bergeron, Y. Canopy Gap Characteristics and Tree Replacement in the Southeastern Boreal Forest. *Ecology* **1998**, *79*, 783–794. [CrossRef]
3. Bergeron, Y.; Harvey, B.; Leduc, A.; Gauthier, S. Forest Management Guidelines Based on Natural Disturbance Dynamics: Stand- and Forest-Level Considerations. *For. Chron.* **1999**, *75*, 49–54. [CrossRef]
4. Pasher, J.; Seed, E.; Duffe, J. Development of Boreal Ecosystem Anthropogenic Disturbance Layers for Canada Based on 2008 to 2010 Landsat Imagery. *Can. J. Remote Sens.* **2013**, *39*, 42–58. [CrossRef]
5. Frank, D.; Finckh, M.; Wirth, C. Impacts of Land Use on Habitat Functions of Old-Growth Forests and their Biodiversity. In *Old-Growth Forests. Ecological Studies (Analysis and Synthesis)*; Wirth, C., Gleixner, G., Heimann, M., Eds.; Springer: Berlin/Heidelberg, Germany, 2009; Volume 207, pp. 429–450. ISBN 978-3-540-92705-1.
6. Grondin, P.; Gauthier, S.; Poirier, V.; Tardif, P.; Boucher, Y.; Bergeron, Y. Have Some Landscapes in the Eastern Canadian Boreal Forest Moved beyond Their Natural Range of Variability? *For. Ecosyst.* **2018**, *5*, 30. [CrossRef]
7. Mosseler, A.; Thompson, I.; Pendrel, B.A. Overview of Old-Growth Forests in Canada from a Science Perspective. *Environ. Rev.* **2003**, *11*, S1–S7. [CrossRef]
8. Stokland, J.; Siitonen, J. Species diversity of saproxylic organisms. In *Biodiversity in Dead Wood*; Stokland, J., Siitonen, J., Jonsson, B., Eds.; Cambridge University Press: Cambridge, UK, 2012; pp. 248–274. ISBN 978-1-139-02584-3.
9. Nilsson, S.G. Selecting biodiversity indicators to set conservation targets: Species, structures, or processes? In *Setting Conservation Targets for Managed Forest Landscapes*; Villard, M.-A., Jonsson, B.G., Eds.; Cambridge University Press: Cambridge, UK, 2009; pp. 79–108. ISBN 978-0-521-87709-1.
10. Boudreault, C.; Gauthier, S.; Bergeron, Y. Epiphytic Lichens and Bryophytes on *Populus Tremuloides* Along a Chronosequence in the Southwestern Boreal Forest of Québec, Canada. *Bryologist* **2000**, *103*, 725–738. [CrossRef]
11. Bergeron, Y.; Fenton, N.J. Boreal Forests of Eastern Canada Revisited: Old Growth, Nonfire Disturbances, Forest Succession, and Biodiversity. *Botany* **2012**, *90*, 509–523. [CrossRef]
12. Martin, M.; Tremblay, J.A.; Ibarzabal, J. An Indicator Species Highlights Continuous Deadwood Supply Is a Key Ecological Attribute of Boreal Old-Growth Forests. *Ecosphere*. under review.
13. Wesołowski, T.; Martin, K. Tree Holes and Hole-Nesting Birds in European and North American Forests. In *Ecology and Conservation of Forest Birds*; Mikusiński, G., Roberge, J.-M., Fuller, R., Eds.; Cambridge University Press: Cambridge, UK, 2018; pp. 79–134. ISBN 978-1-139-68036-3.
14. Drapeau, P.; Nappi, A.; Imbeau, L.; Saint-Germain, M. Standing Deadwood for Keystone Bird Species in the Eastern Boreal Forest: Managing for Snag Dynamics. *For. Chron.* **2009**, *85*, 227–234. [CrossRef]
15. Tremblay, J.A.; Savard, J.-P.L.; Ibarzabal, J. Structural Retention Requirements for a Key Ecosystem Engineer in Conifer-Dominated Stands of a Boreal Managed Landscape in Eastern Canada. *For. Ecol. Manag.* **2015**, *357*, 220–227. [CrossRef]
16. Imbeau, L.; Desrochers, A. Area Sensitivity and Edge Avoidance: The Case of the Three-Toed Woodpecker (*Picoides Tridactylus*) in a Managed Forest. *For. Ecol. Manag.* **2002**, 249–256. [CrossRef]
17. Nappi, A.; Drapeau, P.; Leduc, A. How Important Is Dead Wood for Woodpeckers Foraging in Eastern North American Boreal Forests? *For. Ecol. Manag.* **2015**, *346*, 10–21. [CrossRef]
18. Cadieux, P.; Drapeau, P. Are Old Boreal Forests a Safe Bet for the Conservation of the Avifauna Associated with Decayed Wood in Eastern Canada? *For. Ecol. Manag.* **2017**, *385*, 127–139. [CrossRef]
19. Imbeau, L.; Savard, J.-P.L.; Gagnon, R. Comparing Bird Assemblages in Successional Black Spruce Stands Originating from Fire and Logging. *Can. J. Zool.* **1999**, *77*, 1850–1860. [CrossRef]
20. Imbeau, L.; Mönkkönen, M.; Desrochers, A. Long-Term Effects of Forestry on Birds of the Eastern Canadian Boreal Forests: A Comparison with Fennoscandia. *Conserv. Biol.* **2001**, 1151–1162. [CrossRef]
21. Craig, C.; Mazerolle, M.J.; Taylor, P.D.; Tremblay, J.A.; Villard, M.-A. Predictors of Habitat Use and Nesting Success for Two Sympatric Species of Boreal Woodpeckers in an Unburned, Managed Forest Landscape. *For. Ecol. Manag.* **2019**, *438*, 134–141. [CrossRef]
22. Saucier, J.-P.; Robitaille, A.; Grondin, P.; Bergeron, J.-F.; Gosselin, J. *Les Régions Écologiques Du Québec Méridional*, 4th ed.; Carte à l'échelle de 1/1 250,000; Ministère Des Ressources Naturelles et de La Faune Du Québec: Quebec, QC, Canada, 2011.

23. Saucier, J.-P.; Grondin, P.; Robitaille, A.; Gosselin, J.; Morneau, C.; Richard, P.J.H.; Brisson, J.; Sirois, L.; Leduc, A.; Morin, H.; et al. Écologie forestière. In *Manuel de Foresterie—Nouvelle Édition Entièrement Revue et Augmentée*; Ordre des ingénieurs forestiers du Québec, Ed.; Éditions Mulitmondes: Quebec, QC, Canada, 2009; pp. 165–316.
24. Boucher, Y.; Grondin, P.; Noël, J.; Hotte, D.; Blouin, J.; Roy, G. *Classification Des Écosystèmes et Caractérisation Des Forêts Mûres et Surannées: Le Cas Du Projet Pilote de La Réserve Faunique Des Laurentides*; Gouvernement du Québec, Ministère des Ressources naturelles et de la Faune, Direction de la Recherche Forestière: Quebec, QC, Canada, 2008.
25. Boucher, Y.; Bouchard, M.; Grondin, P.; Tardif, P. *Le Registre Des États de Référence: Intégration Des Connaissances Sur La Structure, La Composition et La Dynamique Des Paysages Forestiers Naturels Du Québec Méridional*; Gouvernement du Québec, Ministère des Ressources naturelles et de la Faune, Direction de la recherche forestière: Quebec, QC, Canada, 2011.
26. Desrochers, A.; Drolet, B. Le Programme de Surveillance Des Oiseaux Nicheurs de La Forêt Montmorency: Une Nouvelle Source de Tendances Des Populations d'oiseaux Nicheurs Pour La Forêt Boréale Au Québec. *Nat. Can.* **2017**, *141*, 61–74. [CrossRef]
27. Boucher, Y.; Grondin, P. Impact of Logging and Natural Stand-Replacing Disturbances on High-Elevation Boreal Landscape Dynamics (1950–2005) in Eastern Canada. *For. Ecol. Manag.* **2012**, *263*, 229–239. [CrossRef]
28. Brandt, J.P. The Extent of the North American Boreal Zone. *Environ. Rev.* **2009**, *17*, 101–161. [CrossRef]
29. Xeno-Canto. Available online: https://www.xeno-canto.org/ (accessed on 1 May 2016).
30. Macaulay Library—The Cornell Lab of Ornithology. Available online: https://www.macaulaylibrary.org/ (accessed on 1 May 2016).
31. Weather Archive in Fort Montmorency, METAR. Available online: https://rp5.ru/Weather_archive_in_Fort_Montmorency,_METAR (accessed on 15 August 2020).
32. Ministère des Forêts, de la Faune et des Parcs. *Norme de Stratification Écoforestière: Quatrième Inventaire Écoforestier du Québec méridional*; Secteur des Forêts. Direction des Inventaires Forestiers: Québec, QC, Canada, 2015; ISBN 978-2-550-73857-2.
33. R Core Team. *R: A Language and Environment for Statistical Computing*; R Foundation for Statistical Computing: Vienna, Austria, 2020.
34. Mackenzie, D.I.; Nichols, J.D.; Lachman, G.B.; Droege, S.; Royle, J.A.; Catherine, A. Langtimm Estimating Site Occupancy Rates When Detection Probabilities Are Less than One. *Ecology* **2002**, 2248–2255. [CrossRef]
35. Fiske, I.; Chandler, R. Unmarked: An R Package for Fitting Hierarchical Models of Wildlife Occurrence and Abundance. *J. Stat. Soft.* **2011**, *43*. [CrossRef]
36. Clement, M.A.; Barrett, K.; Baldwin, R.F. Key Habitat Features Facilitate the Presence of Barred Owls in Developed Landscapes. *ACE* **2019**, *14*, art12. [CrossRef]
37. Burnham, K.P.; Anderson, D.R. *Model Selection and Multi-Model Inference: A Practical Information-Theoretic Approach*, 2nd ed.; Springer: New York, NY, USA, 2002; ISBN 0-387-95364-7.
38. Mazerolle, M.J. AICcmodavg: Model Selection and Multimodel Inference Based on (Q)AIC(c). R Package Version 2.3-0. 2020. Available online: https://repo.bppt.go.id/cran/web/packages/AICcmodavg/vignettes/AICcmodavg.pdf (accessed on 20 May 2020).
39. Schieck, J.; Song, S.J. Changes in Bird Communities throughout Succession Following Fire and Harvest in Boreal Forests of Western North America: Literature Review and Meta-Analyses. *Can. J. For. Res.* **2006**, *36*, 1299–1318. [CrossRef]
40. Imbeau, L.; Desrochers, A. Foraging Ecology and Use of Drumming Trees by Three-Toed Woodpeckers. *J. Wildl. Manag.* **2002**, *66*, 222. [CrossRef]
41. Gagné, C.; Imbeau, L.; Drapeau, P. Anthropogenic Edges: Their Influence on the American Three-Toed Woodpecker (*Picoides dorsalis*) Foraging Behaviour in Managed Boreal Forests of Quebec. *For. Ecol. Manag.* **2007**, *252*, 191–200. [CrossRef]
42. Hagan, J.M.; McKinley, P.S.; Meehan, A.L.; Grove, S.L. Diversity and Abundance of Landbirds in a Northeastern Industrial Forest. *J. Wildl. Manag.* **1997**, *61*, 718. [CrossRef]
43. Vaillancourt, M.-A.; Drapeau, P.; Gauthier, S.; Robert, M. Availability of Standing Trees for Large Cavity-Nesting Birds in the Eastern Boreal Forest of Québec, Canada. *For. Ecol. Manag.* **2008**, *255*, 2272–2285. [CrossRef]
44. Tremblay, J.A.; Leonard, D.L., Jr.; Imbeau, L. American Three-toed Woodpecker (*Picoides dorsalis*). Version 1.0. In *Birds of the World*; Rodewald, P.G., Ed.; Cornell Lab of Ornithology: Ithaca, NY, USA, 2020.
45. Environment Canada. *Recovery Strategy for the Woodland Caribou, Boreal Population (Rangifer Tarandus Caribou) in Canada [Proposed]. Species at Risk Act Recovery Strategy Series*; Environment Canada: Ottawa, ON, Canada, 2011; pp. vi + 55.
46. Halme, P.; Allen, K.A.; Auniņš, A.; Bradshaw, R.H.W.; Brūmelis, G.; Čada, V.; Clear, J.L.; Eriksson, A.-M.; Hannon, G.; Hyvärinen, E.; et al. Challenges of Ecological Restoration: Lessons from Forests in Northern Europe. *Biol. Conserv.* **2013**, *167*, 248–256. [CrossRef]
47. Shorohova, E.; Kneeshaw, D.; Kuuluvainen, T.; Gauthier, S. Variability and Dynamics of Old-Growth Forests in the Circumboreal Zone: Implications for Conservation, Restoration and Management. *Silva Fenn.* **2011**, *45*. [CrossRef]
48. Tikkanen, O.-P.; Martikainen, P.; Hyvärinen, E.; Junninen, K.; Kouki, J. Red-Listed Boreal Forest Species of Finland: Associations with Forest Structure, Tree Species, and Decaying Wood. *Ann. Zool. Fennici* **2006**, *43*, 373–383. [CrossRef]
49. Tingstad, L.; Grytnes, J.A.; Felde, V.A.; Juslén, A.; Hyvärinen, E.; Dahlberg, A. The Potential to Use Documentation in National Red Lists to Characterize Red-Listed Forest Species in Fennoscandia and to Guide Conservation. *Glob. Ecol. Conserv.* **2018**, *15*, e00410. [CrossRef]
50. Haavik, A.; Dale, S. Are Reserves Enough? Value of Protected Areas for Boreal Forest Birds in Southeastern Norway. *Ann. Zool. Fenn.* **2012**, *49*, 69–80. [CrossRef]

51. Nappi, A. Sélection d'habitat et Démographie Du Pic à Dos Noir Dans Les Forêts Brûlées de La Forêt Boréale. Ph.D. Thesis, Université du Québec à Montréal, Montréal, QC, Canada, 2009.
52. Fenton, N.J.; Imbeau, L.; Work, T.; Jacobs, J.; Bescond, H.; Drapeau, P.; Bergeron, Y. Lessons Learned from 12 Years of Ecological Research on Partial Cuts in Black Spruce Forests of Northwestern Québec. *For. Chron.* **2013**, *89*, 350–359. [CrossRef]
53. Tremblay, J.A.; Bélanger, L.; Desponts, M.; Brunet, G. La Restauration Passive Des Sapinières Mixtes de Seconde Venue: Une Alternative Pour La Conservation Des Sapinières Mixtes Anciennes. *Can. J. For. Res.* **2007**, *37*, 825–839. [CrossRef]
54. Desrochers, A. Tendances Ornithologiques du Québec. Available online: https://www.toq.ffgg.ulaval.ca/ (accessed on 13 October 2020).
55. Meehan, T.D.; LeBaron, G.S.; Dale, K.; Michel, N.L.; Verutes, G.M.; Langham, G.M. *Abundance Trends of Birds Wintering in the USA and Canada, from Audubon Christmas Bird Counts, 1966–2017, Version 2.1*; National Audubon Society: New York, NY, USA, 2018.

Article

Implications of Historical and Contemporary Processes on Genetic Differentiation of a Declining Boreal Songbird: The Rusty Blackbird

Robert E. Wilson [1,*], Steven M. Matsuoka [1], Luke L. Powell [2,3,4,5], James A. Johnson [6], Dean W. Demarest [7], Diana Stralberg [8,9] and Sarah A. Sonsthagen [1]

[1] U.S. Geological Survey, Alaska Science Center, Anchorage, AK 99508, USA; smatsuoka@usgs.gov (S.M.M.); ssonsthagen@usgs.gov (S.A.S.)

[2] Institute of Animal Health and Comparative Medicine, University of Glasgow, Glasgow G12 8QQ, UK; luke.l.powell@gmail.com

[3] Department of Biosciences, Durham University, Stockton Road, Durham DH13LE, UK

[4] Biodiversity Initiative, Houghton, MI 49931, USA

[5] Smithsonian Conservation Biology Institute, National Zoological Park, Migratory Bird Center, Washington, DC 20013-7012, USA

[6] United States Fish and Wildlife Service, Migratory Bird Management, Anchorage, AK 99503, USA; jim_a_johnson@fws.gov

[7] United States Fish and Wildlife Service, Division of Migratory Birds, Atlanta, GA 30345, USA; Dean_Demarest@fws.gov

[8] Department of Renewable Resources, University of Alberta, Edmonton, AB T6G 2H1, Canada; diana.stralberg@canada.ca

[9] Natural Resources Canada, Canadian Forest Service, Northern Forestry Centre, Edmonton, AB T6H 3S5, Canada

* Correspondence: robertewilson0289@gmail.com

Citation: Wilson, R.E.; Matsuoka, S.M.; Powell, L.L.; Johnson, J.A.; Demarest, D.W.; Stralberg, D.; Sonsthagen, S.A. Implications of Historical and Contemporary Processes on Genetic Differentiation of a Declining Boreal Songbird: The Rusty Blackbird. *Diversity* 2021, 13, 103. https://doi.org/10.3390/d13030103

Academic Editor: Stacy A. McNulty

Received: 9 February 2021
Accepted: 16 February 2021
Published: 25 February 2021

Publisher's Note: MDPI stays neutral with regard to jurisdictional claims in published maps and institutional affiliations.

Abstract: The arrangement of habitat features via historical or contemporary events can strongly influence genomic and demographic connectivity, and in turn affect levels of genetic diversity and resilience of populations to environmental perturbation. The rusty blackbird (*Euphagus carolinus*) is a forested wetland habitat specialist whose population size has declined sharply (78%) over recent decades. The species breeds across the expansive North American boreal forest region, which contains a mosaic of habitat conditions resulting from active natural disturbance regimes and glacial history. We used landscape genomics to evaluate how past and present landscape features have shaped patterns of genetic diversity and connectivity across the species' breeding range. Based on reduced-representation genomic and mitochondrial DNA, genetic structure followed four broad patterns influenced by both historical and contemporary forces: (1) an east–west partition consistent with vicariance during the last glacial maximum; (2) a potential secondary contact zone between eastern and western lineages at James Bay, Ontario; (3) insular differentiation of birds on Newfoundland; and (4) restricted regional gene flow among locales within western and eastern North America. The presence of genomic structure and therefore restricted dispersal among populations may limit the species' capacity to respond to rapid environmental change.

Keywords: *Euphagus carolinus*; genetic diversity; boreal; glacial refugia; phylogeography

1. Introduction

The spatial organization of suitable habitat across the landscape via historical and contemporary events plays an important role in the maintenance of both genetic and demographic connectivity within plant or animal populations. Discontinuities in habitat, whether from physical barriers or from natural or anthropogenic disturbances, fragment populations into smaller units and can diminish levels of connectivity when (1) the distance between suitable habitat patches exceeds the dispersal capacity of individuals or (2) dispersing individuals do not successfully reproduce. In this way, spatial habitat heterogeneity influences effective dispersal (dispersal followed by reproduction) by individuals, which in turn impacts gene flow and population dynamics (i.e., potential outcome of dispersal, [1,2]). Isolation, whether by distance or environment, often results in lower levels of intra-population genetic variation and higher levels of inter-population genetic differentiation across a species' range, which can have major short- and long-term implications on population persistence. For example, isolated populations with reduced genetic diversity may lose adaptive potential, accumulate deleterious mutations, or experience increased inbreeding [3–7]—all of which can reduce individual fitness and increase vulnerability of populations to decline and extirpation [8–10].

While habitat configuration can have a profound influence on population structure and dynamics, species-specific responses to habitat heterogeneity vary widely due to differences in life history characteristics [11]. In some species, reduced connectivity among habitats may be offset by either high dispersal capabilities [12,13] or the presence of relatively continuous habitat during critical parts of the annual cycle that promote genetic exchange (e.g., mating). Even in highly vagile migratory species, however, habitat fragmentation and availability can contribute to genetic differentiation [14,15]. Thus, the capacity to move may be insufficient alone to offset the negative effects of habitat loss. Effective dispersal (i.e., with reproduction) can have a stronger effect on population demographics than the dispersal event alone [16–18]. Because of this, conservation strategies may need to promote site conditions favorable for successful reproduction in addition to general landscape connectivity, as gene flow is essential for population persistence in a changing environment [16–18]. However, determining the effective outcome of dispersal events is often difficult or nearly impossible with banding or telemetry data alone as neither can directly infer successful reproduction. Genetic signatures can provide insights into the success of natal dispersal and connectivity among breeding areas, which are relevant to the conservation of populations. Population genomics can help identify areas where connectivity across the landscape enriches genetic diversity and enhances the resiliency of populations to environmental perturbation. It can also identify where contemporary or historical limitations in dispersal have led to genomic structuring and distinct populations that may require specific management strategies to remain viable. Thus, population genomics provides a powerful approach to understanding implications of dispersal and can fill information gaps in traditional movement data, especially in migratory birds that nest in remote regions (see references within [15]) such as the vast boreal forest biome of North America.

North America's boreal forest biome contains $\geq$ 25% of the world's wetlands and intact forests [19,20], which provide important habitats to over 300 bird species during the breeding season [21]. While migratory songbirds across North American have undergone concerning population declines, boreal nesting species have exhibited some of the most dramatic declines over the past half century [22,23]. Rusty blackbirds (*Euphagus carolinus*) are an unfortunate example of this pattern; since 1966 the global population size is estimated to have decreased by 78% [23,24], with the decline likely ongoing since the late 19th century [25,26]. Loss of wooded wetlands on the wintering grounds in the southern United States is suspected as a principal driver of these declines [27]. However, methylmercury contamination on the breeding grounds [28–30], conversion of wetland habitats used during migration [31], alteration of boreal wetland breeding habitats, and climate change are also likely contributing factors [27,32,33].

Little is known regarding patterns of genomic connectivity and differentiation among nesting areas of rusty blackbirds, owing to their expansive and remote distribution spanning the boreal forest biome from Alaska eastward to Newfoundland and northern New England. However, banding, migration tracking, and stable-isotope analyses indicate a general migratory divide. Rusty blackbirds nesting in Alaska and western Canada migrate west of the Appalachian Mountains to winter in the Mississippi Alluvial Valley, while birds nesting in eastern Canada and the northeastern United States migrate east of the Appalachian Mountains and winter on the Southeastern Coastal Plain [34–36]. There have been no phylogenetic studies on the species to date, but birds breeding on Newfoundland and Magdelen Islands, Quebec and wintering in South Carolina have been described as a putative subspecies (*E. c. nigrans*) that is phenotypically distinct from the nominate form breeding over the remainder of the range [37].

Here we present a landscape genomic approach to examine the influence of historical (Last Glacial Maximum, LGM) and contemporary processes on patterns of diversity within the rusty blackbird, using reduced representation genomic (double-digest restriction-site associated DNA sequences, ddRAD; biparentally inherited) and mitochondrial DNA data (mtDNA, maternally inherited). Much of the present-day geographic extent of the Nearctic boreal biome was covered by ice sheets during the LGM ~20,000 years ago. Consequently, the post-glacial colonization of the region by flora and fauna expanding their ranges from south of ice sheets or from ice-free glacial refugia such as Alaska (Beringia) and the Atlantic Shelf [38] has influenced how genetic diversity within species is arrayed across boreal landscapes (e.g., [39–42]). Within glacial refugia, populations that diverged in isolation during the LGM generally harbor greater levels of genetic diversity than populations in areas recolonized after glacial retreat, except in areas where lineages from separate refugia intermixed [43]. As the present-day distribution of rusty blackbirds spans the entire North American boreal biome including previously glaciated and unglaciated areas, we hypothesize that (1) rusty blackbirds likely contracted to at least two refugia during the LGM as suggested for other avifauna (e.g., [44]). Further, we hypothesize that (2) the putative subspecies of rusty blackbird (*E. c. nigrans*, [37]) residing on the island of Newfoundland would be differentiated from eastern populations on the mainland through insular isolation, LGM isolation in the Atlantic Shelf refugium, or isolation through other physical barriers (e.g., [45,46]).

Just as the repeated glacial and interglacial cycles of the 2.5 million-year Pleistocene epoch were a powerful force shaping patterns of genetic diversity in boreal forest birds [47], contemporary genetic diversity and structure are also dependent on behavioral and biological aspects of individual species [48] that influence their capacity to move across the landscape [49,50]. Indeed, interpopulation variation in dispersal traits is often related to landscape structure (see [51]). The boreal biome is a mosaic of wetland complexes, upland forests, and montane areas and therefore comprises a naturally fragmented landscape for wetland habitat specialists, such as the rusty blackbird [52]. Although rusty blackbirds are able to move long distances and traverse large gaps in suitable habitat during migration (e.g., Lake Erie, [31], western cordillera, [53]), the broad level migratory connectivity observed [35] may equate to some degree of philopatry within each region (i.e., return to natal area to reproduce). We therefore hypothesize that (3) rusty blackbirds will display genetic structure within western and eastern North America.

2. Materials and Methods

2.1. Sampling and DNA Extraction

Blood was sampled from 205 rusty blackbirds captured from 5 May to 15 July 2009–2018 in mist nets placed near their nests or in post-breeding foraging areas across their breeding range (Figure 1, see Table 1 for sample sizes and locality names). Genomic DNA was extracted using a DNeasy Blood and Tissue kit following the manufacturer's protocols (Qiagen, Valencia, CA, USA). Extractions were quantified using a Broad Range Quant-iT dsDNA Assay Kit (Thermo Fisher Scientific, Inc., Waltham, MA USA).

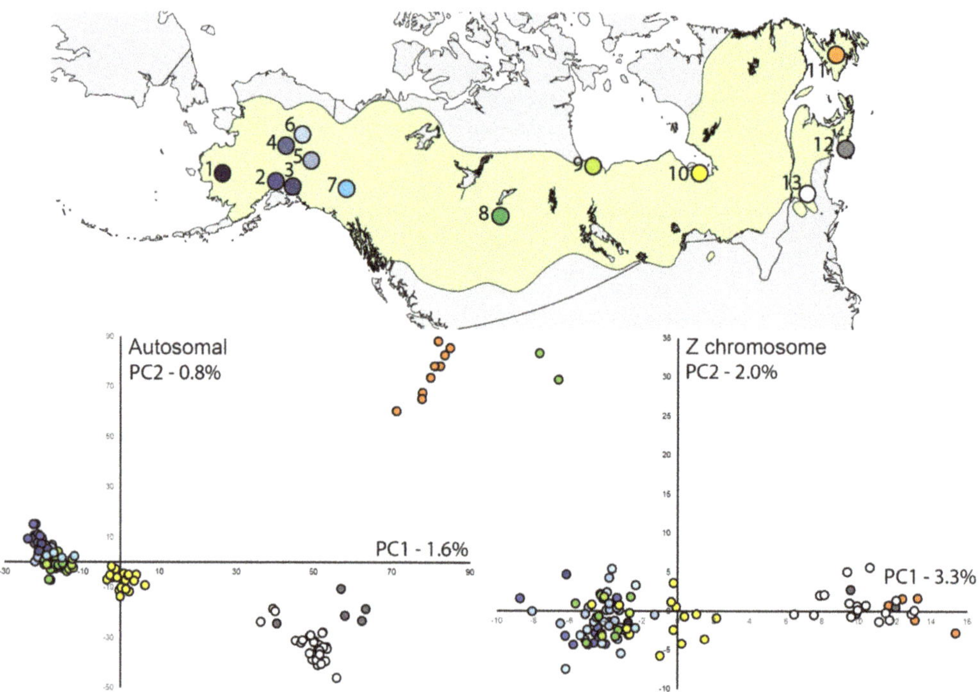

Figure 1. Nesting distribution of rusty blackbirds with sampling locations color-coded and numbered (see Table 1). We note that location 13 (white circle) includes sample locations in Vermont and New Hampshire, USA, and location 12 (grey circle) includes both Nova Scotia and New Brunswick, Canada. The scatter plots are of the first two principal components plotted for haplotypic data, with the proportion of variance explained, from 6205 autosomal and 231 Z-linked loci (males only, because principal components analysis (PCA) does not accommodate heterogamy).

2.2. ddRAD-seq Library Preparation

Sample preparation for ddRAD sequencing followed the double-digest protocol outlined in DaCosta and Sorenson [54]. Genomic DNA (~1 µg) was digested with high fidelity versions of *Sbf*I and *Eco*RI restriction enzymes (New England Biolabs, Ipswich, MA, USA). Amplification and sequencing adapters containing unique barcode or index sequences were ligated to the sticky ends generated by the restriction enzymes. Libraries were size selected using gel electrophoresis (size range 300–450 bp) and purified using a MinElute Gel Extraction Kit (Qiagen) following the manufacturer's protocol. Size-selected fragments were amplified with Phusion high-fidelity DNA polymerase (Thermo Scientific, Pittsburgh, PA, USA) for 20 cycles, and purified using AMPure XP beads (Beckman Coulter, Inc., Indianapolis, IN, USA). Libraries were pooled in equimolar amounts determined via quantitative PCR (KAPA Biosystems, Wilmington, MA, USA). Single-end (150 bp) sequencing was completed on an Illumina HiSeq 4000 at the University of Oregon Core Genomics Facility. Raw Illumina reads are accessioned on National Center for Biotechnology Information (NCBI) Sequence Read Archive (BioProject PRJNA699594, Biosample accessions: SAMN17803887-SAMN17804091, see [55] for additional sample information).

Table 1. Indices of genetic diversity for rusty blackbirds sampled across their North American nesting distribution. Descriptive statistics are listed by marker type (ddRAD autosomal and Z-linked loci, and mtDNA control region) and include nucleotide diversity (π: autosomal [A], z-linked [Z]), effective population size (Ne) based on the molecular co-ancestry method, number of haplotypes (H), haplotype diversity (h) along with sample size (n). Locations are listed west to east. Single standard deviation and 95% confidence limits are in parentheses. Significant values are indicated by asterisks. Refer to Figure 1 for location numbers.

	Location	ddRAD					mtDNA				
		$\pi - A$	$\pi - Z$	Ne	n	H	π	h	Fs	D	n
1	Bethel, Alaska, USA	0.0065	0.0041	–	6	4	0.0046 (0.0033)	0.867 (0.129)	0.3	−0.4	6
2	Anchorage, Alaska, USA	0.0069	0.0048	46.7 (38.5–55.7)	24	13	0.0050 (0.0031)	0.899 (0.046)	−5.1*	−1.2	24
3	Cordova, Alaska, USA	0.0065	0.0045	–	6	2	0.0012 (0.0013)	0.333 (0.215)	1.6	−1.1	6
4	Tanana, Alaska, USA	0.0067	0.0048	–	9	4	0.0022 (0.0017)	0.694 (0.147)	0.0	−0.8	9
5	Tetlin NWR, Alaska, USA	0.0070	0.0049	48.7 (36.6–62.5)	30	14	0.0038 (0.0025)	0.893 (0.040)	−6.3*	−1.4	31
6	Yukon Flats, Alaska, USA	0.0070	0.0049	21.3 (18.4–24.5)	25	15	0.0048 (0.0030)	0.952 (0.029)	−9.2*	−0.8	22
7	Yukon Territory, Canada	0.0058	0.0043	–	4	3	0.0034 (0.0029)	0.833 (0.222)	0.0	1.1	4
8	Alberta, Canada	0.0069	0.0048	26.5 (23.5–29.7)	21	12	0.0039 (0.0025)	0.922 (0.035)	−5.8*	−1.1	22
9	Manitoba, Canada	0.0066	0.0048	–	7	7	0.0063 (0.0042)	1.000 (0.076)	−3.6*	−0.4	7
10	Ontario, Canada	0.0070	0.0048	41.3 (33.9–49.3)	23	15	0.0036 (0.0024)	0.949 (0.028)	−11.6*	−1.2	23
11	Newfoundland, Canada	0.0062	0.0037	19.1 (16.7–21.6)	10	4	0.0040 (0.0027)	0.691 (0.128)	0.9	−0.4	11
12	Nova Scotia/New Brunswick, Canada	0.0061	0.0040	–	5	7	0.0069 (0.0045)	1.000 (0.076)	−3.8*	0.1	7
13	New Hampshire/Vermont, USA	0.0068	0.0046	22.4 (19.4–25.6)	28	13	0.0047 (0.0029)	0.878 (0.041)	−5.2*	−0.3	28
	All populations	0.0061	0.0050	–	198	–	–	–	–	–	200

2.3. Bioinformatics

Illumina reads were demultiplexed at the core facility and processed using the computational pipeline described by DaCosta and Sorenson ([54], custom Python scripts [56]). Briefly, the pipeline filters and clusters reads into putative loci based on sequence similarity (85%) using custom scripts and the UCLUST function in USEARCH v.5 [57]. Genomic positions of loci were determined by BLAST analysis [58] to the *Taeniopygia guttata* (Zebra Finch GenBank assembly reference GCA_003957565.1) reference genome. MUSCLE v.3 [59] was used to align the reads within each cluster with genotyping of samples completed through custom scripts (See [60]). Genotypes were scored homozygous if > 93% of sequence reads were consistent with a single haplotype, whereas heterozygotes were scored if a second haplotype was represented by at least 29% of the reads, or if a second haplotype was represented by 20–29% of reads and the haplotype was present in other individuals. Loci were also "flagged" if the number of single-nucleotide polymorphisms (SNPs) was > 10, and if > 3 SNPs showed strong linkage. Alignments were visually inspected in Geneious (Biomatters Inc. San Francisco, CA), which allowed us to retain loci with insertion/deletions or high levels of polymorphism. To limit any biases due to sequencing error and/or allelic dropout, a minimum of 10 total reads was required to score a genotype as heterozygous with alleles with less than 5X coverage were scored as missing. Loci with a median depth of 10 per individual, < 10% missing genotypes, and < 10% flagged genotypes across all individuals were retained for downstream analyses.

Finally, autosomal and Z chromosome-linked loci were identified as described in Lavretsky et al. [60], with assignments based on differences in sequencing depth and homozygosity between males and females. Chromosomal positions across loci were attained by using blastn results against the *Taeniopygia guttata* reference genome from previous steps in the bioinformatic pipeline. These positions were used to verify marker type based on sequencing depth. Because females have only one Z chromosome, Z-linked markers in females are expected to appear homozygous and be recovered at about half the sequencing depth of males.

2.4. mtDNA Sequencing

Rusty blackbird individuals were sequenced at the mtDNA control region domain I and II. We amplified a 543-base pair (bp) fragment using primer pairs RUBL_CR114L (5′–TCTTTGCCCCCATCAGACAGC–3′) and BBCR_Rev1 [61]. Polymerase chain reaction (PCR) amplifications, cycle-sequencing protocols, and post-sequencing processing followed Sonsthagen et al. [62], with one exception: excess dNTPs and primer were removed using ExoSAP-IT (ThermoFisher Scientific, Waltham, MA, USA). PCR products were sequenced at Functional Biosciences, Inc. (Madison, WI, USA). For quality control purposes, we extracted, amplified, and sequenced 10% of the samples in duplicate. No inconsistencies in mtDNA sequences were observed between replicates. Sequence data are accessioned on GenBank (accession numbers: MW574845-MW574901; see [55] for additional sample information).

2.5. Population Divergence and Nucleotide Diversity

We calculated (1) nucleotide diversity (π) of each ddRAD locus (autosomal and Z-linked) and overall (all loci combined) and (2) composite pairwise estimates of relative divergence (ϕ_{ST}) between sample locations using a custom Python script (out2phistA.py [63]).

Haplotype (h) and nucleotide (π) diversity were calculated for mtDNA in ARLEQUIN 2.0 [64]. Fu's F_S [65] and Tajima's D [66] were calculated to test the hypothesis of selective neutrality and evidence of population fluctuations as implemented in ARLEQUIN. We applied critical significance values of 5%, which requires a p-value < 0.02 for Fu's F_S [65]. An unrooted haplotype network for mtDNA loci was constructed in NETWORK 5.0.1.1 (Fluxus Technology, Suffolk, England 2019) using the reduced median method [67] to illustrate possible reticulations in the gene tree because of homoplasy or recombination. The degree of genetic divergence within rusty blackbirds was assessed by calculating overall and pairwise F_{ST} [frequency-based] and Φ_{ST} using a nucleotide substitution model [68] in ARLEQUIN.

2.6. Population Structure—ddRAD

Population structure was analyzed using four complementary methods with different underlying assumption requirements: (1) principal components analysis (PCA, nonparametric method) to identify major trends in the distribution of genetic variation; (2) maximum likelihood clustering analysis to estimate the number of underlying populations using individual SNPs in the program ADMIXTURE (parametric method); (3) fineRADstructure utilizing haplotypes (concatenation of all variable sites at each locus) to assess contemporary genetic relationships based on shared co-ancestry; and (4) estimate effective migration surfaces (EEMS, [69]) to identify regions that deviate from a null model of isolation-by-distance (IBD).

First, a PCA was implemented on Autosomal and Z-linked loci separately using haplotypic/allelic data and the dudi.pca function in the adegenet R package [70,71]. As PCAs require individuals to be either diploid or haploid, we only included males (in birds, the sex with two copies of Z chromosome) in the analysis of the Z chromosome loci. We plotted individuals relative to the first two principal components to determine the degree that genetically similar individuals cluster into distinct geographic groups.

Second, maximum likelihood estimates of population assignments across individuals were obtained with ADMIXTURE v.1.3 [72,73]. We used all autosomal bi-allelic SNPs with singletons (i.e., rare SNPs observed in only one individual) excluded and without a priori assignment of individuals to populations. First, SNPs were formatted for analyses using plink [74], following steps outlined in Alexander et al. [75]. ADMIXTURE analysis was run with a 10-fold cross-validation, and a quasi-Newton algorithm employed to accelerate convergence [76]. Each analysis used a block relaxation algorithm for point estimation and terminated once the change (i.e., delta) in the log-likelihood of the point estimations increased by < 0.0001. To limit any possible stochastic effects from single analyses, we ran 100 iterations at each population of K (from K of 1–18). The optimum K was based on the average of CV-errors across the 100 analyses per K; however, additional K's were analyzed for further population structure resolution. We then used the program CLUMPP v.1.1 [77] to determine the robustness of the assignments of individuals to populations at each K. First, ADMIXTURE outputs were converted into CLUMPP input files at each K using the R program PopHelper [78]. In CLUMPP, we employed the Large Greedy algorithm and 1000 random permutations to estimate final admixture proportions for each K with per sample assignment probabilities (Q estimates, the log likelihood of group assignment) based on all 100 replicates per K.

Third, we used the fineRADstructure program [79] to cluster individuals into populations with indistinguishable genetic ancestry using a haplotype-based approach. FineRADstructure focuses on the most recent coalescent events (common ancestry) providing information on recent sample relatedness which can be informative regarding levels of contemporary gene flow. Samples were assigned to populations using 5,000,000 iterations sampled every 1000 steps with a burn-in of 500,000. We used 1,000,000 iterations of the tree-building algorithm to assess genetic relationships among clusters. Finally, the output was visualized using the R scripts, fineradstructureplot.r and finestructurelibrary.r [80].

We implemented the spatial method EEMS [69] to estimate effective gene flow (m) and genetic diversity (q) in order to identify areas across the breeding range that deviate from the null expectations of IBD. This method is based on a stepping-stone model where individuals are allowed to move between neighboring demes and gene flow rates can vary by locality. Expected genetic dissimilarity under the model depends on sample location and gene flow rates. Regions where genetic dissimilarity decays more quickly than expected are identified as barriers to gene flow or, conversely, corridors where genetic dissimilarity decayed more slowly than expected. A migration surface that correlates genetic variation with geography is interpolated to visualize potential barriers or corridors to movement. In addition, the model estimates an effective diversity parameter (q), which is the expected within-deme coalescent time and is proportional to average heterozygosity.

We used the same set of SNPs as in Admixture and calculated a dissimilarity matrix using bed2diffs R code included with the EEMS package [81]. An outer coordinate file was constructed using Google Maps API v3 [82] that included the species' entire breeding distribution [52]. Based on preliminary runs, we adjusted parameters, so the accepted proportion of proposals of variance was at least between 10% and 40%. We ran three independent analyses using 10,000,000 burn-in steps followed by 50,000,000 MCMC iterations sampled every 2000 steps for each deme size (100,250). We checked for convergence and visualized effective migration and diversity surfaces using the R package rEEMSplots [69].

2.7. Effective Population Size—ddRAD

Contemporary effective population size (Ne) was estimated from 6184 loci with NeEstimator v.2.1 [83] based on two methods: the linkage disequilibrium (LD) method [84] which tests for the nonrandom associations among alleles at different loci formed by genetic drift in small populations [85] and the molecular co-ancestry method [86] which evaluates the level of allele sharing among individuals. Further, we excluded rare alleles below a range of allele frequency values (Pcrit) from the linkage disequilibrium model to evaluate the effects of low-frequency alleles on Ne estimates. Variance in Ne estimates across a range of Pcrit values suggests a history of gene flow and/or the presence of first-generation dispersers, whereas stable Ne estimates are indicative of isolated populations [87]. We estimated Ne using a haplotype-based approach and Pcrit values between 0.01 and 0.09 and without a frequency restriction. Confidence limits (95%) were determined by jackknifing over loci. Ne was not estimated for sites with a sample size < 10.

2.8. Hindcasted Paleo-Distributions of Breeding Rusty Blackbirds

Finally, we evaluated the paleo-hindcasted maps of the potential LGM breeding range of rusty blackbirds by Stralberg et al. [53]. This was to assess whether population isolation during the LGM may have contributed to genomic structuring detected in our analyses of mtDNA and ddRAD. Models of rusty blackbird breeding density were developed by fitting boosted regression trees [88] to bioclimatic indices derived from current climate normals (1961–1990) as predictors of species abundance data derived from surveys conducted across the species' boreal range [89]. Climate variables were chosen from a set of bioclimatic indices [90] based on several criteria, including relevance to vegetation distributions, avoidance of extreme collinearity, and a preference for seasonal over annual variables when they showed high correlations. The final set of variables included extreme minimum temperature, chilling degree days below 0 °C, growing degree days above 5 °C, seasonal temperature difference, mean summer precipitation, climate moisture index [91], and summer climate moisture index.

Models were then fed inputs from downscaled paleo-climate projections for 21,000 years before present, according to two global climate models, Community Climate Model (CCM1) and Geophysical Fluid Dynamics Laboratory model (GFDL, [92]), to develop hindcast projections of the species' potential LGM breeding distribution [53]. We used projections from Stralberg et al. [53] to develop modified maps of potential LGM density that included areas thought to be covered by the Cordilleran and Laurentide ice sheets. This was to show where suitable habitats may have existed in unglaciated micro-refugia within the major ice sheets or along coastlines adjacent to known refugia now submerged under the sea (e.g., Bering Land Bridge, Grand and Georges Banks; [38]).

3. Results

3.1. Bioinformatics—ddRAD

We obtained over 294,699,715 million raw sequencing reads (median = 1,432,707 reads per individual, range 901,000–1,943,040) with a maximum 150 bp length. Initial exploration of genotyping results revealed that most loci were unambiguously genotyped across samples. We removed seven samples that were deemed to be of close familial relationship (e.g., siblings) based on preliminary PCAs and fineRADstructure results, and field notes (location, date, and age of individuals sampled). For the remaining 198 samples, a total of 6443 clusters (i.e., putative single-copy loci) met the depth/genotype threshold. Of these loci, 6436 passed automated checks for alignment quality or passed thresholds after manual edits that yielded 42,446 SNPs or insertion/deletion (polymorphic sites) from 6381 polymorphic loci. Of those, 6205 loci and 231 loci were assigned to autosomal and the Z chromosome, respectively. Final datasets comprised loci with a median sequencing depth of 118 reads per locus per individual (median range = 73−175 reads/locus/individual), and on average 98.5% (minimum of 80.0%) of alleles per individual per locus were scored.

3.2. Population Divergence and Molecular Diversity

Autosomal nucleotide diversity across the 6205 ddRAD loci was similar for all locations (0.0058–0.0070) with overall value of 0.0061 (Table 1). The highest percentage of loci with no variation (i.e., nucleotide diversity equals zero) was found within Yukon Territory, Canada (21.9%) and Vermont, USA (19.5%). Similar pattern was observed with Z-linked loci with overall nucleotide diversity of 0.0050 (range 0.0037–0.0049, Table 1). Among the 231 Z loci, Yukon Territory (39.4%) and Vermont (38.5%) had the highest percentage of non-variable loci. It should be noted that these populations also had the smallest sample size.

Overall, we uncovered relatively moderate levels of genetic differentiation across sample locations with Z-linked loci showing a 1.5× higher level of differentiation than autosomal (Figure 2; $\phi_{ST}Autosomal$ = 0.019; $\phi_{ST}Z\text{-}linked$ = 0.029). Autosomal and Z-linked loci Φ_{ST} values ranged from 0.002 to 0.082 and −0.016 to 0.144, respectively, with highest degrees of pairwise differentiation found between Newfoundland and other locations and between eastern and western locales (Figure 2). This pattern is reflective in the number of loci showing elevated pairwise divergence (ϕ_{ST} > 0.1) which ranged from 0.2–30.3% of loci (12–1880 loci) across all comparisons (overall 0.8%, n = 52) for autosomal loci and from 0.9–33.3% for Z-linked loci (overall 5.2%, n = 12).

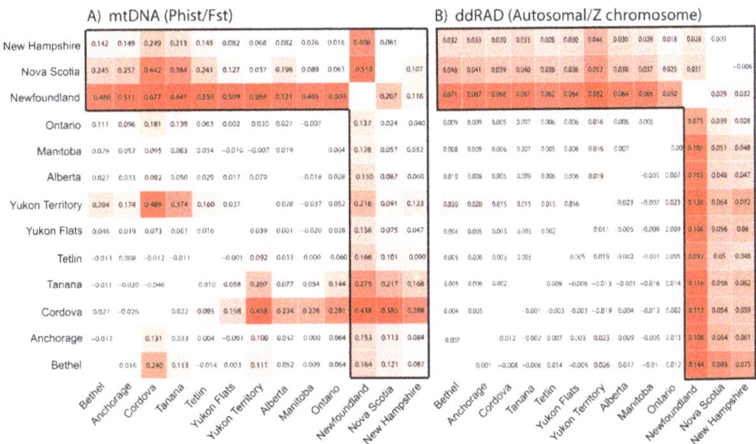

Figure 2. Pairwise Φ_{ST} (above diagonal) and F_{ST} (below diagonal) for mtDNA control region (**A**) and pairwise Φ_{ST} estimate for 6205 autosomal loci (above diagonal) and 231 Z loci (below diagonal (**B**). Darker colors indicate higher values. The northeastern region where higher levels of genetic differentiation were found is outlined.

We uncovered 56 unique mtDNA haplotypes characterized by 36 variable sites. Nucleotide and haplotype diversity were generally similar across sampled locations, except Cordova, Alaska which exhibited lower levels of diversity (n = 4, π = 0.0012, h = 0.333), and Nova Scotia/New Brunswick, Canada (n = 7, π = 0.0069, h = 1.000) and Manitoba (n = 7, π = 0.0063, h = 1.000) which exhibited higher diversity (Table 1). Tajima *D* was not significant, which is consistent with a hypothesis of selective neutrality of mtDNA. Fu's *Fs* was significantly negative for eight populations, suggestive of historical population growth.

Two main haplotype groups were observed in the mtDNA network, although separated by one or three variable sites depending on evolutionary pathway (e.g., presence of reticulations, Figure 3). The first group consisted of mainly Alaska locales with only 2 samples from eastern region and 17 haplotypes though only one haplotype was predominately represented. Conversely, the second group consisted of samples from both regions but was predominately comprised of central and eastern locations and 39 haplotypes with no one dominate haplotype. Genomic structure was uncovered with higher levels of differentiation estimated between Newfoundland and other sample locales, as well as between northeastern and Alaska locales. Similar results were observed with ddRAD loci; pairwise Φ_{ST} values ranged from −0.046 to 0.677 (overall Φ_{ST} = 0.147), and pairwise F_{ST} values ranged −0.014 to 0.438 (overall F_{ST} = 0.073, Figure 2).

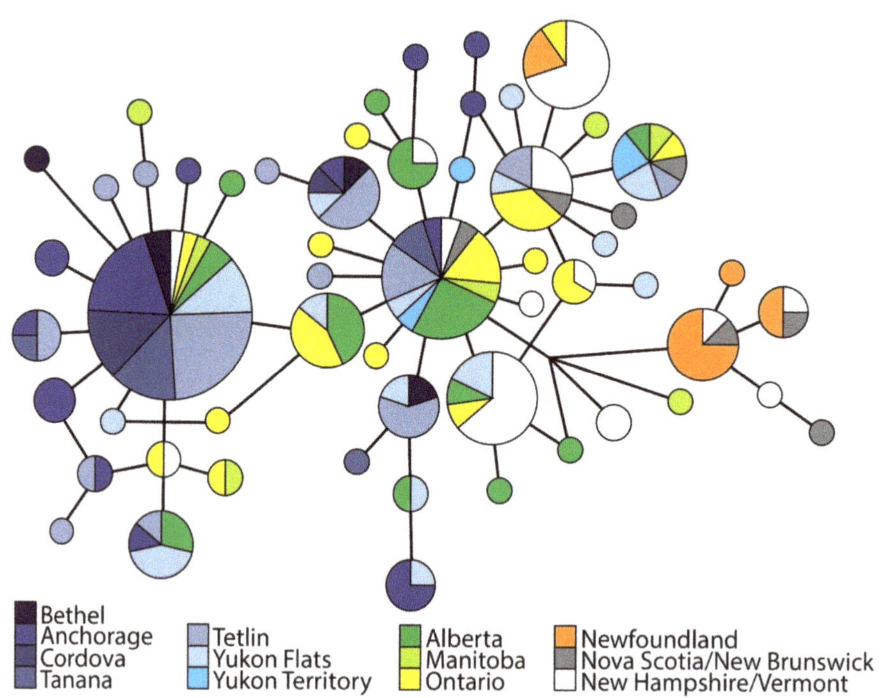

Figure 3. Unrooted mitochondrial DNA haplotype median-joining network for rusty blackbirds. Size of circle is proportional to the frequency of each haplotype observed.

3.3. Population Structure—ddRAD

For PCA, plotting samples relative to the first two principal component axes based on autosomal loci recovered four main clusters that included (1) Newfoundland, (2) Nova Scotia and northeast United States, (3) Ontario, and (4) central Canada and Alaska (Figure 1). These four clusters were also retained but overlapped more when using only Z-linked loci. When PCA included samples only from Cluster 4 (central Canada through Alaska), Anchorage and Canada (Manitoba and Alberta) samples appear slightly differentiated from all other Alaska samples, although there is overlap in PC components (results not shown).

All possible K values were explored across ADMIXTURE analyses (Figure 4A). When K = 2, samples from northeastern locales (Clusters 1 and 2 in PCA) and Alaska and central Canada (PCA Cluster 4) were assigned with high probability to unique clusters. Ontario (PCA Cluster 3) was intermediate between (~70–80% assignment to Alaska cluster) the two main groups. When K = 3 or 4 are considered, the same overall pattern remained except central Canada (Manitoba and Alberta) and Ontario make up a third cluster with variable assignment probability (40–98%, see Supplementary Figure S1).

FineRADstructure revealed more sub-structuring than ADMIXTURE analyses, with individuals clustering mainly by geographic proximity (Figure 4B, Supplementary Figure S2). Locales in northeast North America had the highest shared co-ancestry values with samples being assigned to (1) Newfoundland, (2) Nova Scotia, and (3) northeast United States. Ontario samples were assigned to their own population but in agreement with ADMIXTURE, it shared higher co-ancestry with all other groups/populations indicating connectivity to western (overall higher with central Canada than Alaska) and northeastern locales. Samples from the central and western breeding range were primarily assigned to three main groups: (1) Alaska (excluding most Anchorage samples), (2) Anchorage, and (3) central Canada (Alberta and Manitoba). Unlike northeastern North America and Ontario where groups were mutually exclusive, there was some admixture indicated by individuals being placed in a non-origin group within these central Canada and western populations, for example, Cordova, Alaska and Yukon Territory individuals being grouped with central Canada.

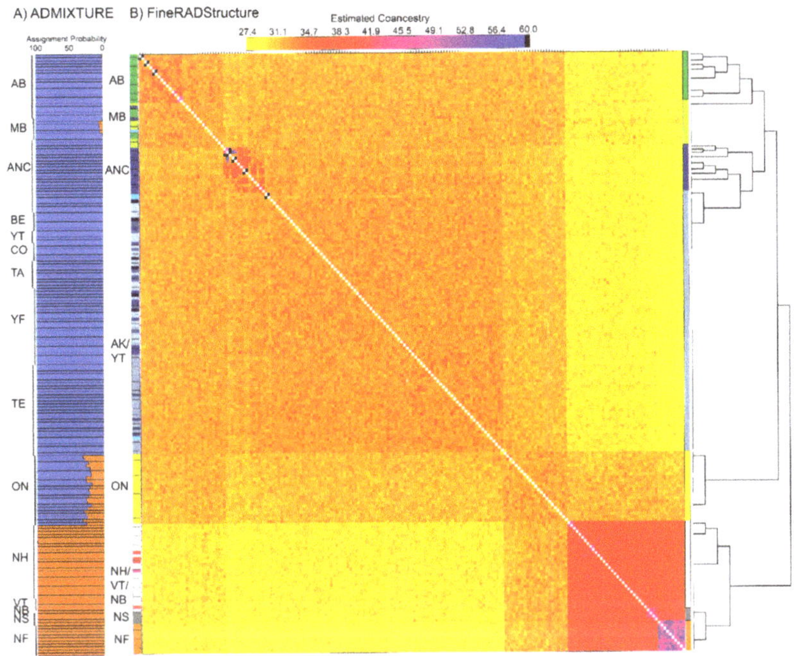

Figure 4. Population assignment analysis using the program (**A**) Admixture and (**B**) fineRADstructure. Average assignment probabilities for K = 2 inferred from bi-allelic single nucleotide polymorphisms in ADMIXTURE. FineRADStructure co-ancestry matrix indicating pairwise genetic similarity between individuals. Inferred populations are indicated by clustering in accompanying dendrogram and locations are indicated by color blocks for general region (solid blocks on right, color was chosen based on locale making up highest proportion of cluster) and by individual (single bars at left). Colors correspond to sampling locations indicated in Figure 1. Co-ancestry values were capped at 60 for illustrative purposes as only a few comparisons were above that value (see Supplementary Figure S2).

EEMS analysis highlighted regions of lower gene flow than expected under IBD. Migration surfaces were similar across deme sizes with one exception. Regions with reduced gene flow were (1) south-central Alaska, (2) Yukon Territory (only present with deme size 250), (3) Alberta, (4) eastern and western Ontario, and (5) Maritime provinces of Canada (Newfoundland, New Brunswick, and Nova Scotia; Figure 5). All these regions had high posterior probabilities (> 0.90) except for the western boundary of Ontario. These boundaries roughly correspond to the population clustering observed in fineRADstructure. High gene flow was characteristic across the remaining distribution.

3.4. Effective Population Size—ddRAD

Although Ne estimates were 10-fold lower for Anchorage, Alaska, and Alberta based on the linkage disequilibrium method, 95% confidence limits overlapped for all of the sampled sites as the upper bounds were infinity (Figure 6). Variation in Ne estimates across Pcrit values was observed for Alaska sites (Tetlin and Yukon Flats), Ontario, and New Hampshire, indicative of past gene flow or the presence of first-generation dispersers affecting Ne estimates. Point estimates for Ne for Newfoundland were infinity, indicating there is no evidence that the population is not very large. However, lower bounds of 95% confidence levels using jackknife method ranged from 83.5 (Pcrit = 0.09) to 372.9

(Pcrit = 0.01) providing a plausible limit for Ne [93]. Conversely, Ne estimates based on the molecular co-ancestry method were lower for two Alaska sites (Anchorage and Tetlin) and Ontario (Table 1).

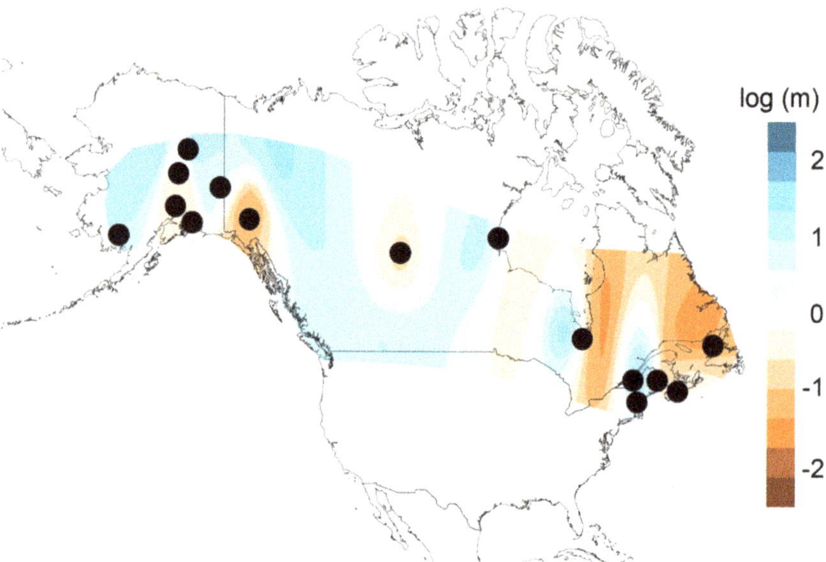

Figure 5. The estimated effective migration surfaces between all sampling locales (black circles) for deme size 250. White areas indicate gene flow rates consistent with isolation by distance expectations whereas shaded areas have dispersal rates that are higher (blue, corridors) or lower (orange, barriers) than expected under isolation-by-distance (IBD). We note that orange area around the Yukon Territory sampling location showed opposite pattern when deme size was lower than 250 (higher than average gene flow rate). For all other areas deme size did not change results.

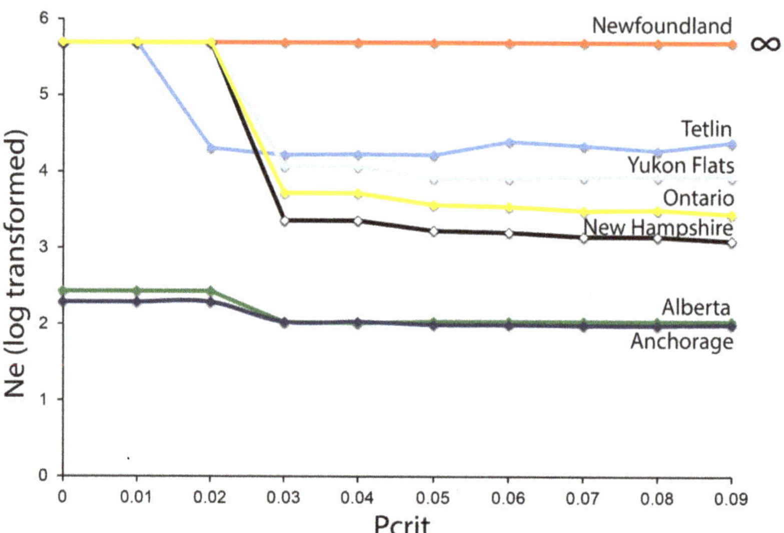

Figure 6. Effective population size (Ne) estimates as a function of excluding rare alleles (Pcrit) in rusty blackbirds sampled across their North American distribution. Point estimates of Ne are log transformed with values >5 signifying infinitely large estimates. Colors correspond to Figure 1.

3.5. Hindcasted Paleo-Distributions of Breeding Rusty Blackbirds

Suitable climate conditions for breeding rusty blackbirds hindcasted to the LGM by Stralberg et al. [53] were located primarily in what is now the central and eastern United States, with some potential suitable habitat in northwestern regions of the United States and Canada (Figure 7). Projected LGM densities were lower, and distributions were more limited and fragmented compared to the species' current distribution and abundance patterns. Despite substantial variation between them, both CCM1 and GFDL models hindcasted suitable climates for paleo-populations of rusty blackbirds in or adjacent to glacial refugia in (1) Alaska and Yukon Territory (Beringia), (2) insular British Columbia (Vancouver Island and Haida Gwaii), (3) Newfoundland (Grand Banks), and (4) New England and the Canadian Maritimes (Georges Banks), as well as south of the ice sheets in (5) British Columbia and Washington and (6) the Great Plains, Upper Midwest, and Northeast United States (Figure 7).

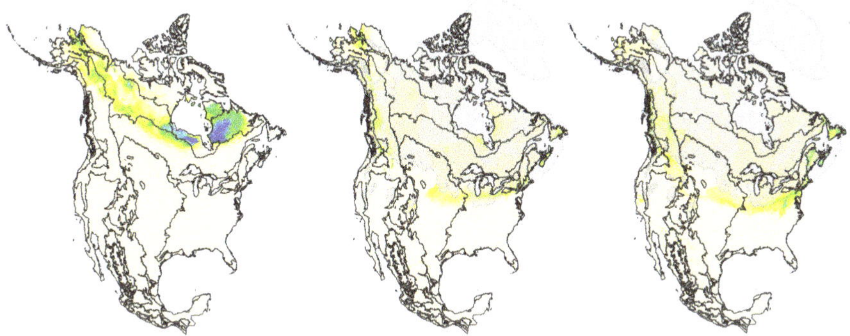

Figure 7. Predictions of current breeding density (**left**) versus hindcasted paleo-breeding densities of rusty blackbirds (based on [53]. Paleo-breeding densities were hindcasted by fitting models of current rusty blackbird density with bioclimatic indices downscaled by Roberts and Hamann [92] for the last glacial maximum (21,000 YBP) using two global climate models: Community Climate Model (CCM1, **middle**) and Geophysical Fluid Dynamics Laboratory (GFDL, **right**). Model-predicted density values range from 0 to 1.4 pairs/ha. Overlaid on hindcast projections are level 1 ecoregion boundaries in black [94] and the extent of Last Glacial Maximum (LGM) ice sheets in transparent gray [95,96].

4. Discussion

The geographic patterns in genomic structure we detected across the rusty blackbird's breeding range conformed to our hypotheses. (1) An east–west partition in genomic structure was observed between rusty blackbirds nesting in the western and central boreal regions versus the eastern boreal region. This was consistent with the east–west migratory divide detected for the species using stable isotopes [35] and suggests long-term separation of populations. (2) As expected, based on insular isolation and plumage differences [37], birds in Newfoundland were differentiated from birds from other sampled sites. (3) Populations within both eastern and western regions exhibited subtle genomic structuring and restricted gene flow, indicating dispersal is limited by discontinuities in habitat, physical barriers, philopatry, or migratory behavior. Further, Ontario appears to be an area of secondary contact between birds originating from eastern and western lineages identified in ADMIXTURE and fineRADstructure analyses. Together, these results indicate that historical and contemporary processes are shaping the distribution of genomic variation among populations of rusty blackbirds across their boreal distribution.

4.1. Pleistocene Influences on Patterns of Genomic Diversity

While the species' current nesting distribution is largely contiguous across the boreal forest biome, the distribution hindcasted to 21,000 years before present was displaced south (other than portions of Beringian Alaska) and was fairly discontinuous and limited (Figure 7, [53]). During the LGM, the glacial ice sheets covered most of northern North America, except Alaska and areas along Pacific and Atlantic coasts [97], and may have sundered the rusty blackbird's nesting range, isolating populations in separate western and eastern refugia, and promoting the partition in genomic variation we detected in multiple analyses of mtDNA and ddRAD (autosomal and Z-linked) loci. In the same way, glacial vicariance is hypothesized to have led to Pleistocene speciation in several sister pairs of temperate conifer boreal bird species [47,98,99]. In addition, several passerines with trans-boreal distributions share this east–west divide in genomic diversity with secondary contact between divergent lineages occurring in the central boreal: mixing occurs from Alberta to Manitoba within the yellow warbler (*Setophaga petechia*, [100,101]), from Alberta to Ontario within Wilson's warbler (*Cardellina pusilla*, [15,102]), from Saskatchewan to Manitoba within the Canada jay (*Perisoreus canadensis*, [41,44]), from Alberta to Ontario within the golden-crowned kinglet (*Regulus satrapa*, [103]), and from Manitoba to Ontario within the rusty blackbird (this study). However, the east–west divide within the blackpoll warbler (*Setophaga striata*) was attributed to isolation by distance and not glacial vicariance [46]. These concordant breaks in genomic diversity across multiple trans-boreal species emphasize the strong influence that the Pleistocene ice sheets played in shaping how genomic variation is arrayed across northern North America in boreal avifauna.

Models of the paleo-breeding distribution indicated that the nesting habitat for rusty blackbirds could have been present in four potential glacial refugia: (1) Alaska (Beringia), (2) Atlantic Shelf, and south of the ice sheets in (3) western (Cordilleran) United States, and (4) eastern (Laurentide) United States, with the eastern region likely a core area based on the relatively high densities inferred during the LGM [53]. These four regions coincide with the locations of glacial refugia proposed for the boreal chickadee [42], the Canada jay [41,44], and the black spruce (*Picea mariana*, [104])—the tree species most often selected for nesting by rusty blackbirds [105]. Rusty blackbirds and other spruce-associated species may have therefore followed the post-glacial colonization routes inferred for black spruce from pollen, fossils, genetics, and ecological modeling [41,106,107]. The spatial apportionment of genomic diversity does suggest that rusty blackbirds occupied at least two refugia during the LGM, although it is not clear from model hindcasts which regions were the refugia nor from the genetic data which sample locations represent refugial populations. Recent colonization of deglaciated areas, whether via long-distance dispersal or leading-edge expansion from glacial refugia, leaves predictable signatures, notably reduced genomic diversity associated with founder events followed by founders preventing subsequent waves of colonizers [108]. Levels of genomic diversity, however, were similar across the rusty blackbird distribution (Table 1). Dispersal likely continued from refugial populations into founding populations in deglaciated areas either via short movements or continued long-distance dispersal. Connectivity between refugial and founding populations would have maintained genomic diversity because effective population sizes would not have been markedly reduced [39,109], thereby erasing the genetic legacy of founder events associated with post-glacial colonization. Further, the eastern clade had high haplotype diversity with no single dominate haplotype suggesting that Newfoundland, Canada Maritime provinces, and New England states were colonized by rusty blackbirds originating from the "core" eastern refugium and possibly the Atlantic Shelf refugium as the effective population size would need to be large to retain and maintain genetic diversity through the LGM. Conversely, the mtDNA network for the western clade had a star-like pattern with a single dominant haplotype predominately represented by Alaska birds. This suggests that the western breeding range from Alaska to Manitoba was colonized by a small refugial population of rusty blackbirds expanding their range from a western or Alaska (Beringia) refugium that supported lower nesting density, and therefore presumably lower effective population size during the LGM.

4.2. Isolation in Newfoundland

Rusty blackbirds occupying Newfoundland were genetically divergent from all sampled locations across marker types. The presence of genetic structure is concordant with the putative subspecies, *Euphagus carolinus nigrans*, described by Burleigh and Peters [37] to nest on Newfoundland and Magdelen Islands, winter in South Carolina, and have a distinctive plumage that is darker, glossier, and bluer than the nominate form. Burleigh and Peters [37] described an additional 12 endemic subspecies of passerines breeding on Newfoundland that also had darker plumages than their mainland counterparts. Similar to rusty blackbird, several of these species and others have since been found to have genetically distinct populations on Newfoundland: the American redstart (*Setophaga ruticilla*, [45]), the blackpoll warbler [46], the boreal chickadee [42], the purple finch (*Haemorhous purpureus*, [110]), the Canada jay [44], the gray-cheeked thrush [111], and the black-capped chickadee (*Poecile atricapillus*, [112]). Cabot Strait ($\geq$104-km wide) and the Strait of Belle Isle ($\geq$15-km wide) separating Newfoundland from mainland Canada therefore seemingly act as strong physical barriers to dispersal by the rusty blackbird and several other passerines.

While Newfoundland rusty blackbirds were genetically differentiated, they were peripheral on the mtDNA network and shared several mtDNA haplotypes with birds breeding elsewhere in eastern North America. This coupled with the limited structure we observed in ADMIXTURE and other analyses of ddRAD loci indicate that divergence between Newfoundland and other eastern locales likely arose post-Pleistocene and has been maintained through restricted dispersal. The observation of shallow divergence is consistent with several other passerines that nest on Newfoundland [44,46,110]. In contrast, other species show much deeper genetic divisions between birds in nesting areas along the northern Atlantic coast versus nearby locales [40,42,112]. In these species, it is more plausible that Newfoundland populations were isolated during the Pleistocene on the Atlantic Shelf refugium offshore of Newfoundland along the Grand Banks [38,97]. Although hindcasting models indicate that suitable habitat was available in Newfoundland during the LGM, the Laurentide ice sheet covered most of the region. Therefore, if rusty blackbirds currently occupying Newfoundland were present in the Atlantic Shelf refugium (Grand Banks, [38,97]), individuals were likely restricted into small isolated area(s) with low numbers. As genetic drift is a strong force shaping genomic diversity in small isolated populations, the lack of deep partitions in genomic variation suggests rusty blackbirds colonized Newfoundland post-Pleistocene and that genomic and morphological variation likely evolved recently as shown for the widely distributed and phenotypically diverse *Junco* species complex [113].

4.3. Contemporary Influences on Patterns of Genomic Variation

We also found several lines of evidence that contemporary processes are limiting dispersal of rusty blackbird populations. First, the maintenance of two distinct genetic groups in western versus eastern North America is indicative of continued restrictions to dispersal between regional nesting areas. This genetic divide mirrored the general migratory divide between western and eastern flyways identified with stable isotopes [35]. Thus, differences in migratory pathways or timing of migration may reinforce reproductive isolation or spatial segregation of regional rusty blackbird populations in the same way as suggested for other passerines with northern distributions [114–117]. We also identified a potential area of secondary contact between eastern and western lineages of rusty blackbirds from a single sample location at the northern border between Ontario and Quebec which had similar levels of recently shared ancestry with both western and eastern regions. Genetic samples coupled with tracking studies from additional eastern locales would help determine the geographic extent of the mixing zone and the degree that migratory behavior may be restricting genomic connectivity [115].

Second, we found evidence of restricted dispersal within eastern and western North America in the form of subtle structure in ddRAD loci among most sampled locales (e.g., Nova Scotia versus other northeastern locales and Alberta/Manitoba versus Alaska) that deviated from expectations based on IBD on a regional scale (Figure 5). Furthermore, three areas (Anchorage, Alberta, and Newfoundland) had similar estimates of effective population size (Ne) across a range of rare alleles, suggesting genomic diversity in these populations is not influenced by ongoing gene flow from adjacent populations. While mountain ranges and ocean bodies are obvious barriers isolating rusty blackbirds in Anchorage and Newfoundland, respectively, there are no clear physical barriers restricting dispersal between Alberta and Manitoba within central Canada. Despite the lack of physical barriers, Alberta individuals were assigned to a non-origin grouping (24%, 5/21) less often than Anchorage individuals (38%, 9/24) in the fineRADstructure analysis, suggesting that other factors are impeding dispersal. Indeed, rusty blackbird nest in boreal forest wetlands that are often distributed in discrete patches separated by unsuitable upland habitat or fragmented by frequent natural disturbance [104,118]. Although rusty blackbirds as well as other passerines migrate over long distances, many species return to near their natal sites to nest where local landscape features influence movements to locate new nesting areas [119]. Other forest dependent birds have shown a reluctance to disperse across large gaps of non-forested habitat to locate new nesting grounds [14,51,120,121]. This type of behavior, if exhibited by rusty blackbirds, may limit dispersal and contribute to genomic structuring among some rusty blackbird nesting areas.

5. Conclusions

Across the breeding range, rusty blackbirds exhibited genomic structuring evident of restricted gene flow, which may limit the species' adaptive capacity to respond to rapid environmental change. The North American boreal biome is a mosaic of wetland complexes and forests that are projected to be transformed as the Earth's climate continues to warm and increase the frequency and magnitude of boreal disturbances such as drought, permafrost thaw, fire, and insect outbreaks [122]. Under various simulations of climate-mediated ecological change over the 21st century, the boreal biome is projected to contract by up to 42% [123], and boreal birds are projected to both dramatically shift their ranges northwards and upwards in elevation and suffer disproportionately high losses in population size and range extent among North America avifauna [22,89,122,124,125]. The rusty blackbird is particularly vulnerable to projected reductions in suitable breeding habitat, which could result in the loss of more than half of the species' breeding range [125] and population numbers [89]. These future declines will exacerbate the species' already steep global population decline and southern range retraction since the mid-20th century [25,26]—the latter already linked to regional trends in warming [33]. Additional research on genotype-environment associations using functionally relevant loci (e.g., transcriptome or gene expression analysis) can build off of the foundation of this study to identify breeding areas that may be more vulnerable to stochastic events as well as areas that pose high conservation value for the species as the climate continues to change (e.g., [101,126]).

The rusty blackbird has an immense migratory range (breeding across the continental boreal biome, wintering over the eastern half of United States) and the many stressors suspected to be contributing to its decline are hypothesized to vary widely across breeding areas and over the annual cycle [26,27,127]. Efforts to understand the causes of decline and efficiently link conservation across this species' annual cycle will therefore benefit from a more comprehensive knowledge of migratory connectivity than the general east–west migratory divide identified through stable isotope and tracking studies [34–36]. Our study is a foundational step in gaining this knowledge as it provides a basis for researchers to infer the natal origins of birds sampled at key migration stopover sites and important wintering areas (e.g., [15]). Understanding migratory connectivity across the rusty blackbird's non-breeding range would, for example, allow researchers to weigh the relative contributions of summer versus winter environmental change on vital rates and population trends (e.g., [128,129]) and enable wildlife managers to strategically target habitat restorations throughout the annual cycle for genetically distinct populations [130]. As the boreal avifauna is among the most rapidly declining groups of birds in North America [23], the integration of information on connectivity with the science and management of recovering rusty blackbird populations could serve as a model for how to restore other poorly studied declining boreal species [27].

Supplementary Materials: The following are available online at https://www.mdpi.com/1424-2818/13/3/103/s1, Figure S1: Admixture results for population clustering, Figure S2: FineRADStructure co-ancestry matrix.

Author Contributions: Conceptualization, R.E.W., S.A.S., L.L.P., S.M.M., J.A.J., D.W.D.; resources, S.M.M., J.A.J., L.L.P., D.W.D.; methodology, R.E.W., S.A.S.; formal analysis, R.E.W., S.A.S., D.S.; investigation, R.E.W., S.A.S.; writing—original draft, R.E.W., S.A.S.; writing—review and editing, R.E.W., S.A.S., L.L.P., S.M.M., J.A.J., D.S., D.W.D.; funding Acquisition, J.A.J., D.W.D., S.A.S.; project administration, S.A.S., L.L.P., S.M.M., J.A.J., D.W.D.; data curation, R.E.W., S.A.S.; fieldwork, J.A.J., L.L.P., S.M.M., All authors have read and agreed to the submitted version of manuscript.

Funding: This research was funded by the U.S. Fish and Wildlife Service, Division of Migratory Bird Management and U.S. Geological Survey, Wildlife Program of the Ecosystems Mission Area.

Institutional Review Board Statement: Sample collections were approved by Institutional Animal Care and Use Committees of the U.S. Fish and Wildlife Service, Alaska Department of Fish and Game (Alaska: IACUC #2008004, approved May 2008) and University of Alberta (Alberta: #AUP00001523).

Informed Consent Statement: Not applicable.

Data Availability Statement: The data used in the present study are accessioned in Sonsthagen et al. [55], NCBI Sequence Read Archive (BioProject PRJNA699594, Biosample accessions: SAMN17803887-SAMN17804091) and NCBI nucleotide database (MW574845-MW574901).

Acknowledgments: We thank Erin Bayne, Paulson Des Brisay, Matthew Brooks, Terry Chesser, Andrew Curtis, Niels Dau, Lucas DeCicco, David Evers, Carol Foss, April Harding-Scurr, David Loomis, Bruce Rodrigues, Pam Sinclair, David Shaw, Marian Snively, David Tessler, and James Wright for their critical contributions to sample collection. Sample collections in Alaska were approved by Institutional Animal Care and Use Committees of the U.S. Fish and Wildlife Service and Alaska Department of Fish and Game. This research used resources provided by the Core Science Analytics, Synthesis, and Libraries (CSASL) Advanced Research Computing (ARC) group at the U.S. Geological Survey. Any use of trade, product or firm names is for descriptive purposes only and does not imply endorsement by the US Government. The findings and conclusions in this article are those of the authors and do not necessarily represent the views of the USFWS. This article has been peer reviewed and approved for publication consistent with USGS Fundamental Science Practices.

Conflicts of Interest: The authors declare no conflict of interest.

References

1. Johnson, A.R.; Wiens, J.A.; Milne, B.T.; Crist, T.O. Animal movements and population dynamics in heterogeneous landscapes. *Landsc. Ecol.* **1992**, *7*, 63–75. [CrossRef]
2. Robbins, A.M.; Robbins, M.M. Fitness consequences of dispersal decisions for male mountain gorillas (*Gorilla beringei beringei*). *Behav. Ecol. Sociobiol.* **2005**, *58*, 295–309. [CrossRef]
3. Frankham, R.; Ballou, J.D.; Briscoe, D.A. *Introduction to Conservation Genetics*; Cambridge University Press: Cambridge, UK, 2002.
4. Keyghobadi, N. The genetic implications of habitat fragmentation for animals. *Can. J. Zool.* **2007**, *85*, 1049–1064. [CrossRef]
5. Radespiel, U.; Bruford, M.W. Fragmentation genetics of rainforests animals: Insights from recent studies. *Conserv. Genet* **2014**, *15*, 245–260. [CrossRef]
6. Oretgo, J.; Aguiree, M.P.; Noguerales, V.; Cordero, P.J. Consequences of extensive habitat fragmentation in landscape-level patterns of genetic diversity and structure in the Mediterranean esparto grasshopper. *Evol. Appl.* **2015**, *8*, 621–632. [CrossRef] [PubMed]
7. de Villemeruil, P.; Rutschmann, A.; Lee, K.D.; Ewen, J.G.; Brekke, P.; Santure, A.W. Little adaptive potential in a threatened passerine bird. *Curr. Biol.* **2019**, *29*, 889–894. [CrossRef]
8. Ouborg, N.J.; Vergeer, P.; Mix, C. The rough edges of the conservation genetics paradigm for plants. *J. Ecol.* **2006**, *94*, 1233–1248. [CrossRef]
9. Leonardi, S.; Piovani, P.; Scalfi, M.; Piotti, A.; Giannini, R.; Menozzi, P. Effect of habitat fragmentation on the genetic diversity and structure of peripheral populations of beech in central Italy. *J. Hered.* **2012**, *103*, 408–417. [CrossRef] [PubMed]
10. Stevens, K.; Harrisson, K.A.; Hogan, F.E.; Cooke, R.; Clarke, R.H. Reduced gene flow in a vulnerable species reflects two centuries of habitat loss and fragmentation. *Ecosphere* **2018**, *9*, e02114. [CrossRef]

11. Ewers, R.M.; Didham, R.K. Confounding factors in the detection of species responses to habitat fragmentation. *Biol. Rev.* **2006**, *81*, 117–142. [CrossRef] [PubMed]

12. Amos, J.N.; Bennett, A.F.; Mac Nally, R.; Newell, G.; Pavlova, A.; Radford, J.Q.; Thomson, J.R.; White, M.; Sunnucks, P. Predicting landscape-genetic consequences of habitat loss, fragmentation and mobility for multiple species of woodland birds. *PLoS ONE* **2012**, *7*, e30888. [CrossRef]

13. Dharmarajan, G.; Beasley, J.C.; Fike, J.A.; Rhodes, O.E., Jr. Effects of landscape, demographic and behavioral factors on kin structure: Testing ecological predictions in a mesopredator with high dispersal capability. *Anim. Conserv.* **2014**, *17*, 225–234. [CrossRef]

14. Lindsay, D.L.; Barr, K.R.; Lance, R.F.; Tweddale, S.A.; Hayden, T.J.; Leberg, P.L. Habitat fragmentation and genetic diversity of an endangered, migratory songbird, the golden-cheeked warbler (*Dendroica chrysoparia*). *Mol. Ecol.* **2008**, *17*, 2122–2133. [CrossRef]

15. Ruegg, K.C.; Anderson, E.C.; Paxton, K.L.; Apkenas, V.; Lao, S.; Siegel, R.B.; DeSante, D.F.; Moore, F.; Smith, T.B. Mapping migration in a songbird using high-resolution genetic markers. *Mol. Ecol.* **2014**, *23*, 5726–5739. [CrossRef]

16. Robertson, E.P.; Fletcher, R.J., Jr.; Cattau, C.E.; Udell, B.J.; Reichert, B.E.; Austin, J.D.; Valle, D. Isolating the roles of movement and reproduction on effective connectivity alters conservation priorities for an endangered bird. *Proc. Natl. Acad. Sci. USA* **2018**, *115*, 8591–8596. [CrossRef]

17. Robertson, E.P.; Fletcher, R.J., Jr.; Austin, J.D.; Valle, D. The number of breeders explains genetic connectivity in an endangered bird. *Mol. Ecol.* **2019**, *28*, 2746–2756. [PubMed]

18. Coulon, A.; Fitzpatrick, J.W.; Bowman, R.; Lovette, J.J. Effects of habitat fragmentation on effective dispersal of Florida scrub jays. *Conserv. Biol.* **2019**, *24*, 1080–1088. [CrossRef]

19. Schindler, D.W.; Lee, P.G. Comprehensive conservation planning to protect biodiversity and ecosystem services in Canadian boreal regions under a warming climate and increasing exploitation. *Biol. Conserv.* **2010**, *143*, 1571–1586. [CrossRef]

20. Wells, J.V. Chapter 1: Threats and conservation status. In *Boreal Birds of North America: A Hemispheric View of Their Conservation Links and Significance*; Studies in Avian Biology; Wells, J.V., Ed.; University of California Press: Berkeley, CA, USA, 2011; Volume 41, pp. 1–6.

21. Wells, J.V.; Blancher, P.J. Global role for sustaining bird populations. In *Boreal Birds of North America: A Hemispheric View of Their Conservation Links and Significance*; Studies in Avian Biology; Wells, J.V., Ed.; University of California Press: Berkeley, CA, USA, 2011; Volume 41, pp. 7–22.

22. Wells, J.; Stralberg, D.; Childs, D. *Boreal Forest Refuge: Conserving North America's Bird Nursery in the Face of Climate Change*; Boreal Songbird Initiative: Seattle, WA, USA, 2018.

23. Rosenberg, K.V.; Dokter, A.M.; Blancher, P.J.; Sauer, J.R.; Smith, A.C.; Smith, P.A.; Stanton, J.C.; Panjabi, A.; Helft, L.; Parr, M.; et al. Decline of the North American avifauna. *Science* **2019**, *366*, 120–124. [CrossRef]

24. Soykan, C.U.; Sauer, J.; Schuetz, J.G.; LeBaron, G.S.; Dale, K.; Langham, G.M. Population trends for North American winter birds based on hierarchical models. *Ecosphere* **2016**, *7*, e01351. [CrossRef]

25. Greenberg, R.; Droege, S. On the decline of the rusty blackbird and the use of ornithological literature to document long-term population trends. *Conserv. Biol.* **1999**, *13*, 553–559. [CrossRef]

26. Greenberg, R.; Matsuoka, S.M. Rusty blackbird: Mysteries of a species in decline (*Euphagus carolinus*). *Condor* **2010**, *112*, 770–777. [CrossRef]

27. Greenberg, R.; Demarest, D.W.; Matsuoka, S.M.; Mettke-Hofmann, C.; Evers, D.; Hamel, P.B.; Luscier, J.; Powell, L.L.; Shaw, D.; Avery, M.L.; et al. Understanding declines in rusty blackbirds. In *Boreal Birds of North America: A Hemispheric View of Their Conservation Links and Significance*; Studies in Avian Biology; Wells, J.V., Ed.; University of California Press: Berkeley, CA, USA, 2011; Volume 41, pp. 107–126.

28. Edmonds, S.T.; Evers, D.C.; Cristol, D.A.; Mettke-Hofmann, C.; Powell, L.L.; McGann, A.J.; Armiger, J.W.; Lane, O.P.; Tessler, D.F.; Newell, P.; et al. Geographic and seasonal variation in mercury exposure of the declining rusty blackbird. *Condor* **2010**, *112*, 789–799. [CrossRef]

29. Edmonds, S.T.; O'Driscoll, N.J.; Hillier, N.K.; Atwood, J.L.; Evers, D.C. Factors regulating the bioavailability of methylmercury to breeding rusty blackbirds in northeastern wetlands. *Environ. Pollut.* **2012**, *171*, 148–154. [CrossRef]

30. Perkins, M.; Lane, O.P.; Evers, D.C.; Sauer, A.; Adams, E.M.; O'Driscoll, N.J.; Edmunds, S.T.; Jackson, A.K.; Hagelin, J.C.; Trimble, J.; et al. Historical patterns in mercury exposure for North American songbirds. *Ecotoxicology* **2020**, *29*, 1161–1173. [CrossRef]

31. Wright, J.R.; Powell, L.L.; Tonra, C.M. Automated telemetry reveals staging behavior in a declining migratory passerine. *Auk Ornithol. Adv.* **2018**, *135*, 461–476. [CrossRef]

32. Powell, L.L.; Hodgman, T.P.; Glanz, W.E.; Osenton, J.D.; Fisher, C.M. Nest-site selection and nest survival of the rusty blackbird: Does timber management adjacent to wetlands create ecological traps? *Condor* **2010**, *112*, 800–809. [CrossRef]

33. McClure, C.J.W.; Rolek, B.W.; McDonald, K.; Hill, G.E. Climate change and the decline of a once common bird. *Ecol. Evol.* **2012**, *2*, 370–378. [CrossRef]

34. Hamel, P.B.; De Stevens, D.; Leininger, T.; Wilson, R. Historical trends in rusty blackbird nonbreeding habitat in forested wetlands. In Proceedings of the 4th International Partners in Flight Conference, McAllen, TX, USA, 13–16 February 2008; Rich, T., Arizmendi, C., Thompson, C., Demarest, D., Eds.; Partners in Flight, 2009; pp. 341–353.

35. Hobson, K.A.; Greenberg, R.; Van Wilgenburg, S.L.; Mettke-Hofmann, C. Migratory connectivity in the rusty blackbird: Isotopic evidence from feathers of historical and contemporary specimens. *Condor* **2010**, *112*, 778–788. [CrossRef]

36. Johnson, J.A.; Matsuoka, S.M.; Tessler, D.F.; Greenberg, R.; Fox, J.W. Identifying migratory pathways used by rusty blackbirds breeding in southcentral Alaska. *Wilson J. Ornithol.* **2012**, *124*, 698–703. [CrossRef]
37. Burleigh, T.D.; Peters, H.S. Geographic variation in Newfoundland birds. *Proc. Biol. Soc. Wash.* **1948**, *61*, 111–126.
38. Dyke, A.S. Late Quaternary vegetation history of northern North America based on pollen, macrofossil, and faunal remains. *Geogr. Phys. Quatern.* **2005**, *59*, 211–262.
39. Shafer, A.B.A.; Cullingham, C.I.; Cote, S.D.; Coltman, D.W. Of glaciers and refugia: A decade of study sheds new light on the phylogeography of northwestern North America. *Mol. Ecol.* **2010**, *19*, 4589–4621. [CrossRef]
40. Sonsthagen, S.A.; Talbot, S.L.; Scribner, K.T.; McCracken, K.G. Multilocus phylogeography and population structure of common eiders breeding in North America and Scandinavia. *J. Biogeogr.* **2011**, *38*, 1368–1380. [CrossRef]
41. van Els, P.; Cicero, C.; Klicka, J. High latitudes and high genetic diversity: Phylogeography of a widespread boreal bird, the gray jay (*Perisoreus canadensis*). *Mol. Phylogenet. Evol.* **2012**, *63*, 456–465. [CrossRef] [PubMed]
42. Lait, L.A.; Burg, T.M. When east meets west: Population structure of high-latitude resident species, the boreal chickadee (*Poecile hudsonicus*). *Heredity* **2013**, *111*, 321–329. [CrossRef] [PubMed]
43. Hewitt, G.M. Genetic consequences of climatic oscillations in the Quaternary. *Philos. Trans. R. Soc. B Biol. Sci.* **2004**, *359*, 183–195. [CrossRef]
44. Dohms, K.M.; Graham, B.A.; Burg, T.M. Multilocus genetic analysis and spatial modeling reveal complex population structure and history in a widespread resident North American passerine (*Perisoreus canadensis*). *Ecol. Evol.* **2017**, *7*, 9869–9889. [CrossRef]
45. Colbeck, G.J.; Gibbs, H.L.; Marra, P.P.; Hobson, K.; Webster, M.S. Phylogeography of a widespread North American migratory songbird (*Setophaga ruticilla*). *J. Hered.* **2008**, *99*, 453–463. [CrossRef]
46. Ralston, J.; Kirchman, J.J. Continent-scale genetic structure in a boreal forest migrant, the blackpoll warbler (*Setophaga striata*). *Auk* **2012**, *129*, 467–478.
47. Lovette, I.J. Glacial cycles and the tempo of avian speciation. *Trends Ecol. Evol.* **2005**, *20*, 57–59. [CrossRef]
48. Zamudio, K.R.; Bell, R.C.; Mason, N.A. Phenotypes in phylogeography: Species' traits, environmental variation, and vertebrate diversification. *Proc. Natl. Acad. Sci. USA* **2016**, *113*, 8041–8048. [CrossRef] [PubMed]
49. Cote, J.; Bestion, E.; Jacob, S.; Travis, J.; Legrand, D.; Baguette, M. Evolution of dispersal strategies and dispersal syndromes in fragmented landscapes. *Ecography* **2017**, *40*, 56–73. [CrossRef]
50. Cayuela, H.; Rougemont, Q.; Prunier, J.G.; Soore, J.-S.; Clobert, J.; Besnard, A.; Bernatchez, L. Demographic and genetic approaches to study dispersal in wild animal populations: A methodological review. *Mol. Ecol.* **2018**, *27*, 3976–4010. [CrossRef]
51. Cornelius, C.; Awade, M.; Cândia-Gallardo, C.; Sieving, K.E.; Metzger, J.P. Habitat fragmentation drives inter-population variation in dispersal behavior in a Neotropical rainforest bird. *Perspect. Ecol. Conserv.* **2017**, *15*, 3–9. [CrossRef]
52. Avery, M.L. Rusty blackbird (*Euphagus carolinus*), Version 1.0. In *Birds of the World*; Poole, A.F., Ed.; Cornell Lab of Ornithology: Ithaca, NY, USA, 2020.
53. Stralberg, D.; Matsuoka, S.M.; Handel, C.M.; Bayne, E.M.; Schmiegelow, F.K.A.; Hamman, A. Biogeography of boreal passerine range dynamics in western North America: Past, present, and future. *Ecography* **2017**, *40*, 1050–1066. [CrossRef]
54. DaCosta, J.M.; Sorenson, M.D. Amplification biases and consistent recovery of loci in a double-digest RAD-seq protocol. *PLoS ONE* **2014**, *9*, e106713. [CrossRef] [PubMed]
55. Sonsthagen, S.A.; Wilson, R.E.; Matsuoka, S.M.; Johnson, J.A.; Demarest, D.W.; Stralberg, D.; Powell, L.L. Rusty blackbird (*Euphagus carolinus*) genetic data, North America. U.S. Geological Survey Data Release. 2021. [CrossRef]
56. BU-RAD-seq ddRAD-seq Pipeline. Available online: http://github.com/BU-RAD-seq/ddRAD-seq-Pipeline (accessed on 30 June 2020).
57. Edgar, R.C. Search and clustering orders of magnitude faster than BLAST. *Bioinformatics* **2010**, *26*, 2460–2461. [CrossRef]
58. Altschul, S.F.; Gish, W.; Miller, W.; Myers, W.E.; Lipman, D.J. Basic local alignment search tool. *J. Mol. Biol.* **1990**, *2015*, 403–410. [CrossRef]
59. Edgar, R.C. MUSCLE: Multiple sequence alignment with high accuracy and high throughput. *Nucleic Acids Res.* **2004**, *32*, 1792–1797. [CrossRef] [PubMed]
60. Lavretsky, P.; DaCosta, J.M.; Hernández-Baños, B.E.; Engils, A., Jr.; Sorenson, M.D.; Peters, J.L. Speciation genomics and a role for the Z chromosome in the early stages of divergence between Mexican ducks and mallards. *Mol. Ecol.* **2015**, *25*, 661–674. [CrossRef]
61. Dufort, M.J.; Barker, F.K. Range dynamics, rather than convergent selection, explain the mosaic distribution of red-winged blackbird phenotypes. *Ecol. Evol.* **2013**, *3*, 4910–4924. [CrossRef]
62. Sonsthagen, S.A.; Talbot, S.L.; McCracken, K.G. Genetic characterization of common eiders (*Somateria mollissima*) breeding on the Yukon-Kuskokwim Delta, Alaska. *Condor* **2007**, *109*, 879–894. [CrossRef]
63. BU-RAD-Seq Out-Conversions. Available online: http://github.com/BU-RAD-seq/Out-Conversions (accessed on 30 June 2020).
64. Schneider, S.; Roessli, D.; Excoffier, L. *ARLEQUIN Version 2.0: A Software for Population Genetic Analyses*; Genetics and Biometry Laboratory, University of Geneva: Geneva, Switzerland, 2000.
65. Fu, Y.X. Statistical tests of neutrality of mutations against population growth, hitchhiking and background selections. *Genetics* **1997**, *147*, 915–925. [CrossRef]
66. Tajima, F. The effect of change in population size on DNA polymorphism. *Genetics* **1989**, *123*, 597–601. [CrossRef]

67. Bandelt, H.J.; Forster, P.; Sykes, B.C.; Richards, M.B. Mitochondrial portraits of human populations. *Genetics* **1995**, *152*, 743–753. [CrossRef]

68. Tamura, K.; Nei, M. Estimation of the number of nucleotide substitutions in the control region of the mitochondrial DNA in humans and chimpanzees. *Mol. Biol. Evol.* **1993**, *10*, 512–526. [PubMed]

69. Petkova, D.; Novembre, J.; Stephens, M. Visualizing spatial population structure with estimated effective migration surfaces. *Nat. Genet.* **2016**, *48*, 94–100. [CrossRef]

70. Dray, S.; Dufour, A.-B. The ade4 package: Implementing the duality diagram for ecologists. *J. Stat. Softw.* **2007**, *22*, 1–20. [CrossRef]

71. Jombart, T. adegent: An R package for the multivariate analysis of genetic markers. *Bioinfomatics* **2008**, *24*, 1403–1405. [CrossRef]

72. Alexander, D.H.; Novembre, J.; Lange, K. Fast model-based estimation of ancestry in unrelated individuals. *Genome Res.* **2009**, *19*, 1655–1664. [CrossRef] [PubMed]

73. Alexander, D.H.; Lange, K. Enhancements to the ADMIXTURE algorithm for individual ancestry estimation. *BMC Bioinfomatics* **2011**, *12*, 246. [CrossRef]

74. Purcell, S.; Neale, B.; Todd-Brown, K.; Thomas, L.; Ferreira, M.A.; Bender, D.; Maller, J.; Sklar, P.; de Bakker, P.I.W.; Daly, M.J.; et al. PLINK: A toolset for whole-genome association and population-based linkage analysis. *Am. J. Hum. Genet.* **2007**, *81*, 559–575. [CrossRef]

75. Alexander, D.H.; Novembre, J.; Lange, K. *Admixture 1.22 Software Manual*; Cold Spring Harbor Lab: New York, NY, USA, 2012.

76. Zhou, H.; Alexander, D.; Lange, K. A quasi-Newton acceleration for high-dimensional optimization algorithms. *Stat. Comput.* **2011**, *21*, 261–273. [CrossRef] [PubMed]

77. Jakobsson, M.; Rosenberg, N.A. CLUMPP: A cluster matching and permutation program for dealing with label switching and multimodality in analysis of population structure. *Bioinformatics* **2007**, *23*, 1801–1806. [CrossRef]

78. Francis, R.M. Pophelper: An R package and web app to analyse and visualize population structure. *Mol. Ecol. Resour.* **2017**, *17*, 27–32. [CrossRef]

79. Malinksy, M.; Trucchi, E.; Lawson, D.J.; Falush, D. RADpainter and fineRADstructure: Population inference from RADseq data. *Mol. Biol. Evol.* **2018**, *35*, 1284–1290.

80. fineRADstructure. Available online: http://cichlid.gurdon.cam.ac.uk/fineRADstructure.html (accessed on 30 June 2020).

81. EEMS. Available online: https://github.com/dipetkov/eems (accessed on 30 June 2020).

82. Google Maps API v3 Tool. Available online: http://www.birdtheme.org/useful/v3tool.html (accessed on 30 June 2020).

83. Do, C.; Waples, R.S.; Peel, D.; Macbeth, G.M.; Tillett, B.J.; Ovenden, J.R. NeEstimator v2: Re-implementation of software for the estimation of contemporary effective population size (Ne) from genetic data. *Mol. Ecol. Resour.* **2014**, *14*, 209–214. [CrossRef]

84. Waples, R.S.; Do, C. LDNE: A program for estimating effective population size from data on linkage disequilibrium. *Mol. Ecol. Resour.* **2008**, *8*, 753–756. [CrossRef]

85. Luikart, G.; Ryman, N.; Tallmon, D.A.; Schwartz, M.K.; Allendorf, F.W. Estimation of census and effective population sizes: The increasing usefulness of DNA-based approaches. *Conserv. Genet.* **2010**, *11*, 355–373. [CrossRef]

86. Nomura, T. Estimation of effective number breeders from molecular coancestry of single cohort sample. *Evol. Appl.* **2008**, *1*, 462–474. [CrossRef]

87. Waples, R.S.; England, P.R. Estimating contemporary effective population size on the basis of linkage disequilibrium in the face of migration. *Genetics* **2011**, *189*, 633–644. [CrossRef]

88. Elith, J.; Leathwick, J.R.; Hastie, T. A working guide to boosted regression trees. *J. Anim. Ecol.* **2008**, *77*, 802–813. [CrossRef] [PubMed]

89. Stralberg, D.; Matsuoka, S.M.; Hamann, A.; Bayne, E.M.; Sólymos, P.; Schmiegelow, F.K.A.; Wang, X.; Cumming, S.G.; Song, S.J. Projecting boreal bird responses to climate change: The signal exceeds the noise. *Ecol. Appl.* **2015**, *25*, 52–69. [CrossRef]

90. Wang, T.; Hamann, A.; Spittlehouse, D.L.; Murdock, T.Q. ClimateWNA—high-resolution spatial climate data for western North America. *J. Appl. Meteorol. Climatol.* **2012**, *51*, 16–29. [CrossRef]

91. Hogg, E.H. Temporal scaling of moisture and the forest-grassland boundary in western Canada. *Agric. For. Meteorol.* **1997**, *84*, 115–122. [CrossRef]

92. Roberts, D.R.; Hamann, A. Predicting potential climate change impacts with bioclimate envelope models: A palaeoecological perspective. *Glob. Ecol. Biogeogr.* **2012**, *21*, 121–133. [CrossRef]

93. Waples, R.S.; Do, C. Linage disequilibrium estimates of contemporary Ne using highly variable genetic markers: A largely untapped resource for applied conservation and evolution. *Evol. Appl.* **2010**, *3*, 244–262. [CrossRef]

94. CEC. *Ecological Regions of North America: Toward a Common Perspective*; Commission for Environmental Cooperation: Montreal, QC, Canada, 1997; 60p.

95. Ehlers, J.; Gibbard, P.L.; Hughes, P.D. (Eds.) *Quaternary Glaciations—Extent and Chronology*; Elsevier: Amsterdam, The Netherlands, 2011; 1108p.

96. LGM Glaciation Extends. Available online: https://crc806db.uni-koeln.de/layer/show/6 (accessed on 18 February 2021).

97. Dyke, A.S. An outline of North American deglaciation with emphasis on central and northern Canada. In *Quaternary Glaciations—Extent and Chronology, Part II: North America, Developments in Quaternary Science*; Ehlers, J., Gibbard, P.L., Eds.; Elsevier: Amsterdam, The Netherlands, 2004; Volume 2b, pp. 373–424.

98. Johnson, N.K.; Cicero, C. New mitochondrial DNA data affirm the importance of Pleistocene speciation in North American birds. *Evolution* **2004**, *58*, 1122–1130. [CrossRef] [PubMed]

99. Weir, J.; Schluter, D. Ice sheets promote speciation in boreal birds. *Proc. R. Soc. Lond. B Biol. Sci.* **2004**, *271*, 1881–1887. [CrossRef]

100. Boulet, M.; Gibbs, H.L.; Hobson, K.A. Integrated analysis of genetic, stable isotope, and banding data reveal migratory connectivity and flyways in the northern yellow warbler (*Dendroica petechia*; Aestiva group). *Ornithol. Monogr.* **2006**, *61*, 29–78. [CrossRef]

101. Bay, R.A.; Harrigan, R.J.; Underwood, V.L.; Gibbs, H.L.; Smith, T.B.; Ruegg, K. Genomic signals of selection predict climate-driving population declines in a migratory bird. *Science* **2018**, *359*, 83–86. [CrossRef]

102. Irwin, D.E.; Irwin, J.H.; Smith, T.B. Genetic variation and seasonal migratory connectivity in Wilson's warblers (*Wilsonia pusilla*): Species-level differences in nuclear DNA between western and eastern populations. *Mol. Ecol.* **2011**, *20*, 3102–3115. [CrossRef] [PubMed]

103. Graham, B.A.; Carpenter, A.M.; Friesen, V.L.; Burg, T.M. A comparison of neutral genetic differentiation and genetic diversity among migratory and resident populations of golden-crowned kinglets (*Regulus satrapa*). *J. Ornithol.* **2020**, *161*, 509–519. [CrossRef]

104. Gérardi, S.; Jaramillo-Correa, J.P.; Beaulieu, J.; Bousquet, J. From glacial refugia to modern populations: New assemblages of organelle genomes generated by differential cytoplasmic gene flow in transcontinental black spruce. *Mol. Ecol.* **2010**, *19*, 5265–5280. [CrossRef]

105. Matsuoka, S.M.; Shaw, D.; Sinclair, P.H.; Johnson, J.A.; Corcoran, R.M.; Dau, N.C.; Meyers, P.M.; Rojek, N.A. Nesting ecology of Rusty Blackbirds in Alaska and Canada. *Condor* **2010**, *112*, 810–824. [CrossRef]

106. Jaramillo-Correa, J.P.; Beaulieu, J.; Bousquet, J. Variation in mitochondrial DNA reveals multiple distant glacial refugia in black spruce (*Picea mariana*), a transcontinental North American conifer. *Mol. Ecol.* **2004**, *13*, 2735–2747. [CrossRef] [PubMed]

107. Roberts, D.R.; Hamann, A. Glacial refugia and modern genetic diversity of 22 western North American tree species. *Proc. R. Soc. B* **2015**, *282*, 2014–2903. [CrossRef] [PubMed]

108. Hewitt, G.M. The genetic legacy of the Quaternary ice ages. *Nature* **2000**, *405*, 907–913. [CrossRef] [PubMed]

109. Hewitt, G.M. Some genetic consequences of ice ages, and their role in divergence and speciation. *Biol. J. Linn. Soc.* **1996**, *58*, 247–276. [CrossRef]

110. Macfarlane, C.B.A.; Natola, L.; Brown, M.W.; Burg, T.M. Population genetic isolation and limited connectivity in the purple finch (*Haemorphous purpureus*). *Ecol. Evol.* **2016**, *6*, 8304–8317. [CrossRef] [PubMed]

111. FitzGerald, A.M.; Whitaker, D.M.; Ralston, J.; Kirchman, J.J.; Warkentin, I.G. Taxonomy and distribution of the imperilled Newfoundland gray-cheeked thrush, *Catharus minimus minimus*. *Avian Conserv. Ecol.* **2017**, *12*, 10. [CrossRef]

112. Hindley, J.; Graham, B.A.; Burg, T.M. Pleistocene glacial cycles and physical barriers influence phylogeographic structure in black-capped chickadees (*Poecile atricapillus*), a widespread North American passerine. *Can. J. Zool.* **2018**, *96*, 1366–1377. [CrossRef]

113. Friis, G.; Aleixandre, P.; Rodríguez-Estrella, R.; Navarro-Sigüenza, A.G.; Milá, B. Rapid postglacial diversification and long-term stasis within the songbird genus Junco: Phylogeographic and phylogenomic evidence. *Mol. Ecol.* **2016**, *25*, 6175–6195. [CrossRef]

114. Ruegg, K. Genetic, morphological, and ecological characterization of a hybrid zone that spans a migratory divide. *Evolution* **2008**, *62*, 452–466. [CrossRef]

115. Delmore, K.E.; Fox, J.W.; Irwin, D.E. Dramatic intraspecific differences in migratory routes, stopover sites and wintering areas, revealed using light-level geolocators. *Proc. R. Soc. B* **2012**, *279*, 4582–4589. [CrossRef] [PubMed]

116. Delmore, K.E.; Irwin, D.E. Hybrid songbirds employ intermediate routes in a migratory divide. *Ecol. Lett.* **2014**, *17*, 1211–1218. [CrossRef]

117. Cohen, E.B.; Rushing, C.R.; Moore, F.R.; Hallworth, M.T.; Hostetler, J.A.; Ramirez, M.G.; Marra, P.P. The strength of migratory connectivity for birds en route to breeding through the Gulf of Mexico. *Ecography* **2019**, *42*, 658–669. [CrossRef]

118. Machtans, C.S.; Van Wilgenburg, S.L.; Armer, L.A.; Hobson, K.A. Retrospective comparison of the occurrence and abundance of rusty blackbird in the Mackenzie Valley, Northwest Territories. *Avian Conserv. Ecol.* **2007**, *2*, 3. [CrossRef]

119. Belisle, M.; Desrochers, A.; Fortin, M.J. Influence of forest cover on the movements of forest birds: A homing experiment. *Ecology* **2001**, *82*, 1893–1904.

120. Matthysen, E.; Adriaensen, F.; Dhondt, A.A. Dispersal of nuthatches, *Sitta europaea*, in a highly fragmented forest habitat. *Oikos* **1995**, *72*, 375–381. [CrossRef]

121. Matthysen, E.; Currie, D. Habitat fragmentation reduces disperser success in juvenile *Sitta europaea*: Evidence from patterns of territory establishment. *Ecography* **1996**, *19*, 67–72. [CrossRef]

122. Stralberg, D.; Berteaux, D.; Drever, C.R.; Drever, M.; Naujokaitis-Lewis, I.; Schmiegelow, F.K.A.; Tremblay, J.A. Conservation planning for boreal birds in a changing climate: A framework for action. *Avian Conserv. Ecol.* **2019**, *14*, 13. [CrossRef]

123. Rehfeldt, G.E.; Crookston, N.L.; Sáenz-Romero, C.; Campbell, E.M. North American vegetation model for land-use planning in a changing climate: A solution to large classification problems. *Ecol. Appl.* **2012**, *22*, 119–141. [CrossRef] [PubMed]

124. Bateman, B.L.; Taylor, L.; Wilsey, C.; Wu, J.; LeBaron, G.S.; Langham, G. Risk to North American birds from climate change-related threats. *Conserv. Sci. Pract.* **2020**, *2*, e243. [CrossRef]

125. Bateman, B.L.; Wilsey, C.; Taylor, L.; Wu, J.; LeBaron, G.S.; Langham, G. North American birds require mitigation and adaptation to reduce vulnerability to climate change. *Conserv. Sci. Pract.* **2020**, *2*, e242. [CrossRef]

126. Fitzpatrick, M.C.; Keller, S.R. Ecological genomics meets community-level modelling of biodiversity: Mapping the genomic landscape of current and future environmental adaptation. *Ecol. Lett.* **2015**, *18*, 1–16. [CrossRef]

127. Greenberg, R.S. Bye bye blackbird: Rusty blackbirds are vanishing from our southern swamps and northern forests. *Zoogoer* **2008**, *37*, 11–15.
128. Rushing, C.S.; Ryder, T.B.; Marra, P.P. Quantifying drivers of population dynamics for a migratory bird throughout the annual cycle. *Proc. R. Soc. B* **2016**, *283*, 20152846. [CrossRef]
129. Rushing, C.S.; Hostetler, J.A.; Sillett, T.S.; Marra, P.P.; Rotenberg, J.A.; Ryder, T.B. Spatial and temporal drivers of avian population dynamics across the annual cycle. *Ecology* **2017**, *98*, 2837–2850. [CrossRef]
130. Williams, C.K.; Castelli, P.M. 2012. A historical perspective of the connectivity between waterfowl research and management. In *Wildlife Science: Connecting Research with Management*; Sands, J.P., Brennan, L.A., DeMaso, S.J., Schnupp, M.J., Eds.; CRC Press: Boca Raton, FL, USA, 2012; pp. 155–178.

Article

Assessment of Rusty Blackbird Habitat Occupancy in the Long Range Mountains of Newfoundland, Canada Using Forest Inventory Data

Kathleen K. E. Manson [1], Jenna P. B. McDermott [2], Luke L. Powell [3,4,5,*], Darroch M. Whitaker [6] and Ian G. Warkentin [1,2,7]

[1] School of Environmental Sciences, University of Guelph, Guelph, ON N1G 2W1, Canada;
    kathleenmanson1@gmail.com (K.K.E.M.); ian.warkentin@grenfell.mun.ca (I.G.W.)
[2] Cognitive and Behavioural Ecology Program, Memorial University of Newfoundland,
    St. John's, NL A1B 3X7, Canada; jmcdermott@mun.ca
[3] Institute of Animal Health and Comparative Medicine, University of Glasgow, Glasgow G12 8QQ, UK
[4] Department of Biosciences, Durham University, Stockton Road, Durham DH13LE, UK
[5] Biodiversity Initiative, Houghton, MI 49931, USA
[6] Parks Canada, Rocky Harbour, NL A0K 4N0, Canada; darroch.whitaker@canada.ca
[7] Environmental Science, Memorial University of Newfoundland, Corner Brook, NL A2H 6P9, Canada
*   Correspondence: Luke.L.Powell@gmail.com

Received: 13 May 2020; Accepted: 29 August 2020; Published: 4 September 2020

Abstract: Rusty blackbirds (*Euphagus carolinus*), once common across their boreal breeding distribution, have undergone steep, range-wide population declines. Newfoundland is home to what has been described as one of just two known subspecies (*E. c. nigrans*) and hosts some of the highest known densities of the species across its extensive breeding range. To contribute to a growing body of literature examining rusty blackbird breeding ecology, we studied habitat occupancy in Western Newfoundland. We conducted 1960 point counts across a systematic survey grid during the 2016 and 2017 breeding seasons, and modeled blackbird occupancy using forest resource inventory data. We also assessed the relationship between the presence of introduced red squirrels (*Tamiasciurus hudsonicus*), an avian nest predator, and blackbird occupancy. We evaluated 31 *a priori* models of blackbird probability of occurrence. Consistent with existing literature, the best predictors of blackbird occupancy were lakes and ponds, streams, rivers, and bogs. Red squirrels did not appear to have a strong influence on blackbird habitat occupancy. We are among the first to model rusty blackbird habitat occupancy using remotely-sensed landcover data; given the widespread availability of forest resource inventory data, this approach may be useful in conservation efforts for this and other rare but widespread boreal species. Given that Newfoundland may be a geographic stronghold for rusty blackbirds, future research should focus on this distinct population.

Keywords: red squirrel; boreal; wetland; *Euphagus carolinus*; point count; remotely sensed landscape data; unmarked

1. Introduction

Though the rate of decline may have eased in the last decade [1], long-term monitoring has documented range-wide declines in rusty blackbird (*Euphagus carolinus*) populations that exceed 80% over the past century, with qualitative evidence for declines dating back to the 19th century [2–4]. Migratory species such as blackbirds can be affected by stressors or threats acting during any or all phases of the annual cycle [5,6]. The factors driving this dramatic range-wide decline in the formerly wide-spread and common rusty blackbird are not fully understood, though this species may have been affected by threats manifested during at least two phases of the annual cycle. Multiple stressors on the species' wintering grounds have likely contributed to declines, including: "Pest" control measures on agricultural land that target other species of blackbirds but also lead to incidental rusty blackbird mortality [7]; loss of up to 80% of potential wintering wetland habitat through conversion to agricultural land use and fragmentation [3]; and exposure to high levels of methyl mercury through dietary intake (during both wintering and breeding) that could cause physiological or reproductive impacts [8]. Factors acting on the species' boreal forest breeding grounds have also likely contributed to population declines [2], including wetland conversion for agricultural purposes [9,10] and habitat degradation linked to climate change [11,12]. Regarding the latter, McClure et al. [11] detected a 143 km northward shift in the southern edge of the breeding range of rusty blackbirds between 1966 and 2005, and also found a significant correlation between rusty blackbird population declines detected in Breeding Bird Survey (BBS) data and Pacific Decadal Oscillations. McClure et al. [11] suggested that warmer temperatures and drier conditions may reduce the amount of arthropod prey and change prey phenology, resulting in a temporal disconnect between breeding phenology and prey availability.

Habitat and habitat quality can play key roles in determining both distribution and productivity of forest songbirds [13,14] and boreal songbirds are known to be influenced by factors that include natural [15,16] and anthropogenic disturbance [17,18]. Broad-scale breeding season habitat associations have been described for rusty blackbirds [4] and various factors have been hypothesized as influencing them on their breeding grounds. These factors include competition with other icterids [19], timber harvesting [20,21], changes to wetland hydrology and ecology [11], and nest predator dynamics [22]. However, though most jurisdictions across the boreal biome maintain relatively standardized forest resource inventory landcover databases to support the management of forests and other natural resources, few studies have quantitatively assessed breeding habitat for rusty blackbirds using forest inventory data. Given their widespread availability and direct role in natural resource management planning, forest inventory data may be particularly useful for efficient, cost-effective monitoring, and for studying a widespread and rare species such as the Rusty Blackbird. This is especially true in the boreal forest, which is often remote or inaccessible to surveyors. Further, these landcover inventories may be used to identify key habitat for conservation and management of species across a wide geographic area, and so may prove useful when applied in the conservation of boreal species at risk.

The breeding range of rusty blackbirds extends across the boreal forest of Canada and the United States, and within this biome breeding activity is most often associated with wetland and riparian ecosystems and adjacent dense conifer stands [4]. During the breeding period, rusty blackbirds forage primarily on aquatic invertebrates along the shorelines of lakes and streams but occasionally seek terrestrial prey [22,23]. Powell et al. [24] assessed breeding site occupancy of rusty blackbirds using ground-based measurement of habitat features (i.e., they did not use remotely sensed data); occupancy was affected by various factors acting at multiple spatial scales, but was driven primarily by the availability of wetlands that afforded suitable foraging opportunities (i.e., areas of shallow water) and evidence of beaver (*Castor canadensis*) activity. The presence of dense patches of conifers in the vicinity of those wetlands was also required for nesting, while stand age and harvest history were less influential [24]. Lack of specificity in the latter matches the largely anecdotal descriptors of nest substrates reported elsewhere; rusty blackbirds appear to prefer nesting in short (less than 4.5m tall), dense conifer stands [21,22,25–27], and predominantly use black spruce (*Picea mariana*) and balsam fir (*Abies balsamea*) near wetlands [24]. They have also been reported nesting in willow thickets (*Salix* sp.; [23]). Most recently, Wohner et al. [27] assessed rusty blackbird habitat use during various periods of the breeding season. They found that streams, softwood and mixed wood sapling stands, wetlands, and areas characterized by slopes between 1% and 8% were important in predicting rusty blackbird occupancy. Streams were very important in predicting nest sites and adult occupancy, but were especially important in predicting fledgling occupancy. In contrast, fledglings and adults strongly selected wetlands, but this habitat was not strongly associated with nest sites. Wohner et al. [27] proposed that streams may be a more important source of smaller arthropod prey for nestlings, whereas wetlands may host larger prey—for example, dragonflies—valuable for adults and large dependent fledglings.

There has been limited research on the impact of nest predators on rusty blackbirds. Matsuoka et al. [25] assessed nesting success in rusty blackbirds and found that of failed nests, 89% were lost to depredation, and various studies have shown that predation risk can affect nest site selection in songbirds (e.g., [28]). In a study of rusty blackbird nest success, Luepold et al. [22] found that North American red squirrels (*Tamiasciurus hudsonicus*) were the most frequent predator of rusty blackbird nests in Maine, USA. Red squirrels are an important predator of nests and fledglings of boreal songbirds that can affect populations and community structure (e.g., [29–31]). Red squirrels were introduced to Newfoundland, Canada during the 1960s and spread rapidly; they are now the most important predator of songbird nests on the island [32–34]. Squirrels have been implicated in the imperilment of two endemic Newfoundland songbird subspecies and represent a novel threat that may affect rusty blackbird nesting success, habitat associations, and abundance on Newfoundland [35,36]. During concurrent research at our study site, McDermott et al. [37] determined that during the 2016 summer season that followed a high mast crop (G. Robineau-Charette and D. Whitaker, unpublished data), red squirrels were detected at 18% of survey points. In contrast, during the 2017 summer season which followed a low mast crop, detections occurred at only 5% of survey points. During both years there was a negative relationship between squirrel detections and elevation with no squirrels detected above 515 m in either year (see [37] for further details). Benkman [38] suggested that red squirrel populations on Newfoundland were more than double those of mainland North American populations.

Forest management has been shown to affect breeding distribution, behavior and success in many species of birds (e.g., [39–41]). By contrast, Powell et al. [24] found that recent logging in adjacent uplands did not feature among the variables retained in the best occupancy models for rusty blackbirds in Northern New England, USA. However, an assessment of nesting success in regenerating, recently harvested stands versus older, established stands at the southern edge of the rusty blackbird breeding range in Maine suggested that recently harvested areas are an ecological trap, with nests in older stands being 2.3 times more likely to fledge young than nests in stands <20 years post-harvest [20]. Conversely, Buckley [26] found that nesting success in managed forest stands was comparable to that for other cup-nesting species, while Wohner et al. [27] suggested that regenerating softwood forests provided dense cover for fledglings to hide from predators. More research is needed to examine the role of forest harvesting at territory and landscape scales on habitat occupancy by rusty blackbirds, and to assess the extent to which such findings from southern portions of the breeding range are applicable in more northerly boreal regions.

As has occurred in continental portions of the rusty blackbird breeding range, the population on the island of Newfoundland experienced a significant decline between 1970 and 2014, as estimated from Breeding Bird Survey (BBS) data (−6.33% per year, 95% credible interval −3.58 to −9.35 based on 23 routes; [1]). Despite this decline, the number of individuals encountered per BBS route on Newfoundland was substantially higher than for all other regions ([7]; mean of 2.03 birds per route on Newfoundland based on data from 1980–2005 compared to a survey-wide mean of 0.26 birds per route for data collected from 1966–2005). As is the case for many bird species found on the island of Newfoundland [42–44], it has been suggested that the rusty blackbird population breeding on Newfoundland and possibly some adjacent portions of Atlantic Canada may be a distinct subspecies (*E. c. nigrans*) from that found across the remainder of the boreal forest (*E. c. carolinus*; [45]), and so may represent a distinct conservation unit for rusty blackbirds. Thus, Newfoundland appears to remain a stronghold for rusty blackbirds and is an important element for range-wide conservation planning.

We used an occupancy modelling approach based on two years of systematic survey data from a 257 km^2 study area in western Newfoundland to assess how rusty blackbird occupancy is influenced by habitat. We used forest resource inventory landcover data derived from aerial photography to measure habitat availability, an uncommon approach for rusty blackbirds (but, see Bale et al. [46], Wohner et al. [27], who used aerial photography-derived habitat data). In addition, we evaluated the influence of red squirrel presence on the probability of blackbird occupancy. We predicted that rusty blackbirds would be associated with wet environments—specifically, waterbodies, watercourses, and bogs—and coniferous stands. Furthermore, we predicted that rusty blackbird occupancy and red squirrel presence would be negatively related.

2. Materials and Methods

2.1. Study Site

We collected data in the Main River and upper Humber River watersheds, located on the eastern slope of the Long Range Mountains of western Newfoundland, Canada (49.75° N, 57.25° W; Figure 1; see also [36,37]). The 257 km^2 study area spans an elevation range from 75 m to 608 m, with elevation increasing from southeast to northwest. Landcover is dominated by wet boreal forest [47] containing a mosaic of mixed and single-species stands dominated by balsam fir or black spruce along with white birch (*Betula papyrifera*), tamarack (*Larix laricina*), and white spruce (*P. glauca*). Much of the mature forest consists of a closed canopy with few, large canopy gaps, and trees at higher elevations tend to have more stunted growth forms [48]. 68% of sites had >25% cover of forest stands older than approximately 30 years of age. In a study in the northern boreal forest of Alaska and Yukon Territory, Viglas et al. [49] found that 30 year old trees had an approximately 50% chance of producing cones, and that this probability increased with age. Thus, a large proportion of our survey area may provide valuable habitat for red squirrels. Qualitative evidence suggests there was a large cone crop in 2015–2016, with a lighter cone crop during 2016–2017 (G. Robineau-Charette and D. Whitaker, unpublished data). Approximately 5% of our survey points were located above 550 m, the approximate altitude of the tree line [48]. Various forms of boreal wetlands and aquatic habitat suitable for rusty blackbirds are widespread across the study area, including bogs, fens, and the shorelines of rivers and lakes, while barrens and other natural openings also make up a proportion of landcover. Overall, landcover within our survey point buffers consisted of an average of 3% lakes, 7% bog, 29% coniferous scrub, and 48% balsam fir- and/or black spruce-dominated forest. Natural disturbances such as wildfire and outbreaks of defoliating insects are uncommon at higher elevations due to climatic conditions, leading to the development of mixed age, old growth fir forests having an abundance of canopy gaps and complex vertical structure [48,50]. Trees at this site have been aged at over 250 years old [48], and around our survey points alone, 27% of points contained 50% or more forest greater than 110 years old. Portions of the study area were harvested by clearcutting between 1990 and 2004 resulting in 19.7% of the study area being cleared in cutblocks ranging from 0.30 ha to 197.4 ha; natural regeneration of balsam fir has followed harvest at these sites. The construction of a 60 m-wide electricity transmission corridor during 2016 and 2017 (Figure 1) created a linear strip of cleared land through the study area. All lands in the study area are provincial public lands (i.e., "Crown lands").

2.2. Field Methods

We collected field data from early June through mid-July of 2016 and 2017 to span the period of peak territorial display and defense for most migratory songbirds in the region, including rusty blackbirds. Systematic surveys were carried out across a grid of points spaced 500 m apart (Figure 1), and for the 2017 season we shifted the grid 250 m north and 250 m east so that survey points fell midway between those sampled the previous year (i.e., a diagonal distance of 354 m from the points sampled the previous year). The total number of surveyed points was 991 during 2016 and 969 in 2017. Solitary observers conducted point counts; data collection included four surveyors during 2016 and five surveyors in 2017 (one individual was common to both seasons). Each surveyor sampled 5–12 adjacent points per day between 05:40 h and 14:30 h. This timeline deviated from standardized avian survey protocols such as Breeding Bird Surveys [51] and was devised as part of a survey using call broadcast originally designed to target gray-cheeked thrush (*Catharus minimus*) and red squirrels (see description below, and [37]). However, 85% of our point counts were conducted before 10am, and the probability of rusty blackbird detection did not vary substantially between hours within our survey period, despite a larger standard error after 13:00. Surveyors recorded wind strength using the Beaufort scale, and stopped field work when high winds (>5 Beaufort scale; 29 km/h) or precipitation/fog impaired visual or auditory detections of songbirds (similar to BBS protocol; [51]). We continued to operate in the presence of light drizzle and fog, as these weather conditions are frequent in this climate, particularly in the morning, and we believe that this approach did not detract from our capacity to detect individuals during surveys. Precipitation was recorded as either absent, fog, drizzle, rain, or snow. Surveyors also recorded cloud cover on a scale from 1–5 (0 = no clouds, 1 < 25% cloud cover, 2 = 26–50%, 3 = 51–75%, 4 = 76–99%, 5 = 100%).

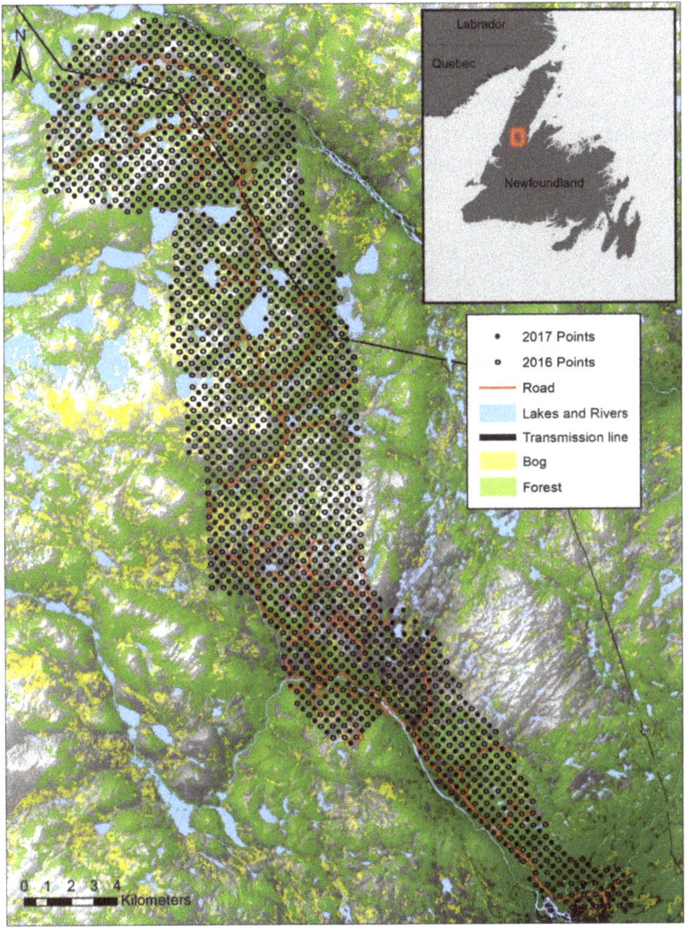

Figure 1. Distribution of survey points in the Main River and upper Humber River watersheds in western Newfoundland, Canada. Each year, points in the survey grid were spaced 500 m apart, and in 2017 the grid was shifted 250 m north and 250 m east, placing the points midway between those sampled the previous year. The location of the study area on the island of Newfoundland is shown by the red box on the inset map.

Surveyors visited each survey point once, conducting an 11-min unlimited radius point count [52] that was divided into the following sequence of five sub-periods: (1) six minutes silent listening; (2) two minutes broadcast of gray-cheeked thrush (*Catharus minimus*) calls and songs (3) a one-minute silent period; (4) one minute broadcast of red squirrel vocalizations; and (5) a final one-minute silent period. These subperiods were designed for an unrelated study that examined the relationship between gray-cheeked thrush and red squirrels. However, we recorded all bird species seen and/or heard and each red squirrel detected during each of the time blocks within the 11 min of a point count. As such, these methods provided suitable data for our study on rusty blackbirds. There is no reason to believe that the broadcast of gray-cheeked thrush vocalizations would influence Rusty Blackbird behavior or detectability. However, rusty blackbirds are known to mob potential predators (e.g., [25,53]) so might be attracted rather than deterred by the squirrel broadcasts. Surveyors used broadcast equipment (FoxPro model FX3 or Crossfire game callers; FoxPro Incorporated, Lewistown, PA 17044, USA) played at a consistent volume; when measured 1 m from the speaker, the peak volume of broadcasts was 82.6 dB.

2.3. Data Analysis

Using ArcMap (version 10.5.1; [54]), we extracted landcover data for each survey point from the provincial forest resource inventory Geographic Information System (GIS) database, which was created using high resolution (sub 10 cm pixel resolution) 3D aerial photographs taken in 2007. Landcover was mapped according to the standard forest resource inventory classification scheme used by the Province of Newfoundland and Labrador, with landcover elements assigned to cover types (e.g., forest, forest scrub, bog, barren, lakes and ponds, rivers). Forest stands were further classified according to 20-year age classes and dominant tree species composition. The provincial forest resource inventory only includes rivers >15 m wide, which are mapped as two-dimensional landcover features (i.e., polygons). However, smaller streams are likely important habitat features for rusty blackbirds [24,27], and are classified as linear features (i.e., 1-dimensional vectors) in Natural Resources Canada's CanVec geospatial database (available under the Government of Canada's open government License [https://open.canada.ca/en]). The national Canvec database is produced using several data sources and resolution varies from 1:10,000–1:50,000 scale. Consequently, we extracted two variables for moving water: (1) The extent of rivers > 15 m wide (m^2) from the provincial forest resource inventory, and (2) the linear length of smaller streams (m) from the CanVec database.

We extracted landcover information within a 347 m radius around each point (i.e., a 37.8 ha circle); this approximates the rusty blackbird home range estimate of 37.5 ± 12.6 ha developed by Powell et al. [20] based on radiotracking 13 rusty blackbirds (6 males, 7 females) in Maine. We converted point count detections of rusty blackbirds into presence-absence data for each point and standardized most habitat features as the proportion of the 37.8 ha buffer circle covered by that habitat type. The only exception was for streams, which were measured as the total length (m) of streams in the 37.8 ha circle, and then re-scaled from 0–1 by dividing these values by the maximum observed stream length (2463.7 m). We assessed these raw landscape variables for collinearity using Spearman's ρ, and did not detect correlations that warranted further consideration or screening of variables (correlation coefficients were less than 0.45, which is below thresholds requiring additional consideration [55]). We aggregated balsam fir and black spruce stands into conifer stands since we believed that they would function similarly as rusty blackbird habitat [4]. Additionally, we excluded habitat features that were present in the landscape but that (1) occurred at less than 10% of the total survey points, or (2) were presumed to be unimportant for rusty blackbirds based on current understanding of the species' habitat needs. The latter habitat features included soil barrens, herbaceous soil barrens, rock barrens, sand, fens, residential land, rights-of-way, cleared land, and forests that were not dominated by either balsam fir or black spruce. Based on this approach our final analysis included seven landcover variables (Table 1).

We used the package UNMARKED [56] in Program R version 3.5.1 [57], using the function "occu" to assess relationships between rusty blackbird occurrence and the seven landcover variables (Table 1) plus year, elevation, and red squirrel presence. This program allowed us to model rusty blackbird detectability prior to running occupancy models. We explored the potential influence of six variables (cloud cover, observer, precipitation type, wind strength, time of day, and day of the year) on likelihood of detection (Table 2) in order to predict occupancy more effectively [58]. Each of the five sub-periods of each point count was considered a site visit, for a total of 5 repeated visits (see [56]). Observer and precipitation type were fit as categorical variables. For both detectability and occupancy modeling (the latter described below), we considered the model having the lowest AICc as the best-fit model and based our conclusions primarily on this model. We then fit 31 *a priori* occupancy models including the null and global models. We used the best-fit detectability model as the base (i.e., null) model for all occupancy models. We formulated these models following the approach of Powell et al. [24] based on information presented in existing rusty blackbird habitat studies [4,24] along with anecdotal reports. Similar to Powell et al. [24], our models included habitats which reflect where one could reasonably expect to find rusty blackbirds. Candidate models included various combinations of landcover variables, including hypothesized interactions between some terms, and this resulted in models containing biologically relevant combinations of (1) nesting habitat, and (2) foraging habitat, and (3) models containing nesting and foraging habitat (Table 3). We also included elevation and red squirrel presence/absence in several of our candidate models. Rather than exploring the influence of geographic coordinates on rusty blackbird occupancy, we considered elevation to be a relevant substitute for assessing overall spatial variation in occupancy. We chose this approach because elevation increases with increased latitude, but east–west dimensions at the study site do not vary drastically. Models having a ΔAICc less than or equal to two were considered to be competing models (i.e., not measurably better than one-another), and the subset of top-ranked models having a cumulative weight of 95% were taken as the best model set [59].

Table 1. Landcover variables used in modeling rusty blackbird occupancy in Newfoundland, Canada, 2016 and 2017. Variables were measured as either the linear amount of the feature (streams [m]) or the proportion of landcover within a 347 m radius of each survey point (all other variables).

Landcover Variable	Description
Lakes and Ponds	Fresh water bodies >0.15 ha at high water mark
Streams	Combined length (m) of all streams <15 m wide
Rivers	Watercourses >15 m wide measured at the high water mark
Bogs	Wetlands where sphagnum moss is the dominant cover type
Conifer	Forest stands where Black Spruce or Balsam Fir makes up 75% or more of the basal area
Coniferous Scrub	Low productivity stands having <10% crown closure and >50% conifer.
Cut90-04	Clearcuts harvested between 1990 and 2004.

Table 2. Candidate models explaining detectability of rusty blackbirds in Newfoundland, Canada, in 2016 and 2017 (n = 1960 survey points).

Model	ΔAIC_c	LL	w_i	K
Observer	0 [a]	−1352.11	0.67	9
Cloud	3.11	−1359.71	0.14	3
Precipitation	4.53	−1357.40	0.07	6
Null	4.76	−1361.53	0.06	3
Wind	6.76	−1361.53	0.02	3
Date	6.83	−1361.56	0.02	3
Time	9.10	−1362.70	0.01	3
Date + Time	11.93	−1363.11	0.00	4

[a] AIC_c of best model = 2722.31.

Table 3. Candidate models describing occupancy (Ψ) of rusty blackbirds in Newfoundland, Canada, in 2016 and 2017 (n = 1960 survey points). Models in the 95% confidence set of best models are highlighted in bold. All models include a term for the effect of observer on detectability.

Model	ΔAICc	LL	wi	K
Ψ Observer~Lakes and Ponds + Bogs + Streams + Rivers	**0** [a]	**−1288.20**	**0.60**	**13**
Ψ Observer~Lakes and Ponds + Bogs + Streams + Conifer Scrub	**2.49**	**−1289.44**	**0.17**	**13**
Ψ Observer~Lakes and Ponds + Bogs + Streams + Conifer	**3.09**	**−1289.74**	**0.13**	**13**
Ψ Observer~Global	**3.74**	**−1284.98**	**0.09**	**18**
Ψ Observer~Lakes and Ponds + Bogs + Cut90-04 + Red Squirrel	20.33	−1298.36	0.00	13
Ψ Observer~Lakes and Ponds + Bogs + Cut90-04	21.04	−1299.73	0.00	12
Ψ Observer~Lakes and Ponds + Bogs	22.66	−1301.55	0.00	11
Ψ Observer~Lakes and Ponds + Bogs + Red Squirrel x Elevation	23.02	−1298.70	0.00	14
Ψ Observer~Lakes and Ponds x Bogs	23.69	−1301.05	0.00	12
Ψ Observer~Lakes and Ponds + Bogs + Conifer + Red Squirrel	23.95	−1300.18	0.00	13
Ψ Observer~Lakes and Ponds + Bogs + Conifer	24.62	−1301.52	0.00	12
Ψ Observer~Lakes and Ponds + Bogs + Rivers + Conifer	24.63	−1300.51	0.00	13
Ψ Observer~Lakes and Ponds + Bogs + Red Squirrel + Elevation	24.63	−1300.51	0.00	13
Ψ Observer~Lakes and Ponds + Bogs + Conifer Scrub + Rivers	24.68	−1300.54	0.00	13
Ψ Observer~Lakes and Ponds + Bogs + Conifer Scrub	24.68	−1301.55	0.00	12
Ψ Observer~Lakes and Ponds + Bogs + Conifer + Conifer Scrub	26.63	−1301.51	0.00	13
Ψ Observer~Lakes and Ponds + Conifer	59.06	−1319.75	0.00	11
Ψ Observer~Bogs + Streams + Conifer + Red Squirrel	64.39	−1320.39	0.00	13
Ψ Observer~Bogs + Streams + Conifer Scrub	65.25	−1321.84	0.00	12
Ψ Observer~Bogs + Streams + Conifer	65.72	−1322.07	0.00	12
Ψ Observer~Lakes and Ponds + Conifer Scrub	68.91	−1324.68	0.00	11
Ψ Observer~Streams + Conifer	76.38	−1328.41	0.00	11
Ψ Observer~Bogs + Conifer + Red Squirrel	77.57	−1328	0.00	12
Ψ Observer~Bogs + Rivers + Conifer + Red Squirrel	78.76	−1327.58	0.00	13
Ψ Observer~Bogs + Conifer	79.37	−1329.91	0.00	11
Ψ Observer~Bogs + Rivers + Conifer	80.79	−1329.61	0.00	12
Ψ Observer~Bogs + Conifer Scrub	81.58	−1331.01	0.00	11
Ψ Observer~Bogs + Rivers + Conifer Scrub	83.27	−1330.85	0.00	12
Ψ Observer~Streams + Conifer Scrub	84.00	−1332.23	0.00	11
Ψ Observer~NULL	119.73	−1352.11	0.00	9
Ψ Observer~Year	121.46	−1351.96	0.00	10

[a] AIC_c of best model = 2602.58.

3. Results

Observers identified rusty blackbirds at 209 of the 1960 points visited over two years (105 points in 2016, 104 points in 2017), a naïve occupancy rate of 10.7%. At 174 sites we detected only one individual, whereas at 30 sites we observed 2 individuals, and at 5 sites we observed 3 individuals. The factor that most strongly affected detectability was observer (Table 2), whereas the model including cloud cover was marginally better than the null model but performed considerably worse than the model only including observer (ΔAICc = 3.11, wi = 0.14). The model including precipitation performed similarly to the null model (ΔAICc = 4.53, wi = 0.07; Table 2). A post-hoc check of an observer + cloud model revealed that adding cloud cover improved model fit only slightly (AICc = 2721.15). Models including wind, time of day, and ordinal day were all worse than the null model, and a post-hoc assessment of the model containing time of day and observer did not prove to be important in detection probability. Based on these findings, we included observer in the base model for all subsequent occupancy models.

Four models were included in our 95% confidence set of best occupancy models, and of these the model that best predicted rusty blackbird occupancy contained lakes and ponds (β = 7.21 ± 0.94), bogs (β = 3.53 ± 0.73), streams (β = 2.06 ± 0.41), and rivers (β = 6.94 ± 3.57) (Table 3). This model included strong positive relationships for lakes and ponds, bogs, and streams, and a weaker positive relationship for rivers (Figure 2a–d). Based on this model, the mean predicted occupancy of rusty blackbirds across our study area was 12.2% (95% confidence interval = 9.4–15.7%; Figure 3). The next two models in the best model set were similar to the best model, but with rivers being replaced by either conifer scrub (β = −0.38 ± 0.44), or conifer forest (β = 0.14 ± 0.36); no strong directional relationship was observed between the probability of rusty blackbird occurrence and conifer scrub cover or conifer forest (Table 3; Figure 2e–f). The final model in our best model set was the global model. Red squirrels were more abundant in the summer of 2016 compared with the summer of 2017 (i.e., they were more abundant following a high cone production year), with 84% of squirrels detected in 2016 [37]. While red squirrel presence did not appear in our 95% confidence set of best occupancy models, adding red squirrel to the best-fit model improved it slightly (AICc = 2601.99; (β = −0.49

± 0.33). Additionally, a red squirrel by year interaction term was added to the best model post hoc; inclusion of this interaction term among the best models was not supported (AICc = 2605.17). We followed the same process with an interaction term that included red squirrel and elevation, and found that this interaction term also did not improve our best model (AICc = 2604.16). We found no evidence that squirrels meaningfully influenced rusty blackbird distribution or occupancy (Table 3; Figure 2 g).

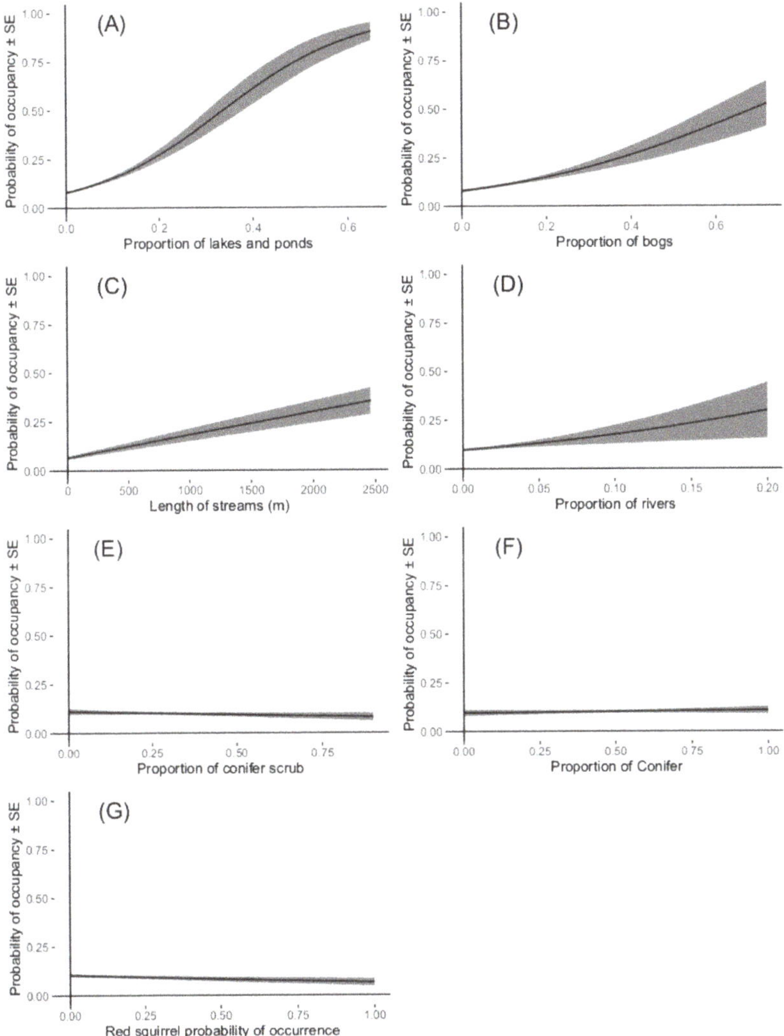

Figure 2. Relationship between rusty blackbird probability of occurrence and landcover variables found in our highest-ranked occupancy models (**A–F**), as well as with red squirrel occurrence (**G**), in Newfoundland, Canada, in 2016 and 2017 (*n* = 1960). Landcover variables were (**A**) proportion of lakes and ponds (**B**) proportion of bogs (**C**) length of stream (**D**) proportion of rivers, (**E**) proportion of conifer scrub, (**F**) proportion of conifer forest, and (**G**) red squirrel probability of occurrence. Each variable was plotted using the best model that included that factor (Table 3). Error bars show one standard error. Note that while the vertical axis is the same for all plots, the scale and units of the horizontal axis varies.

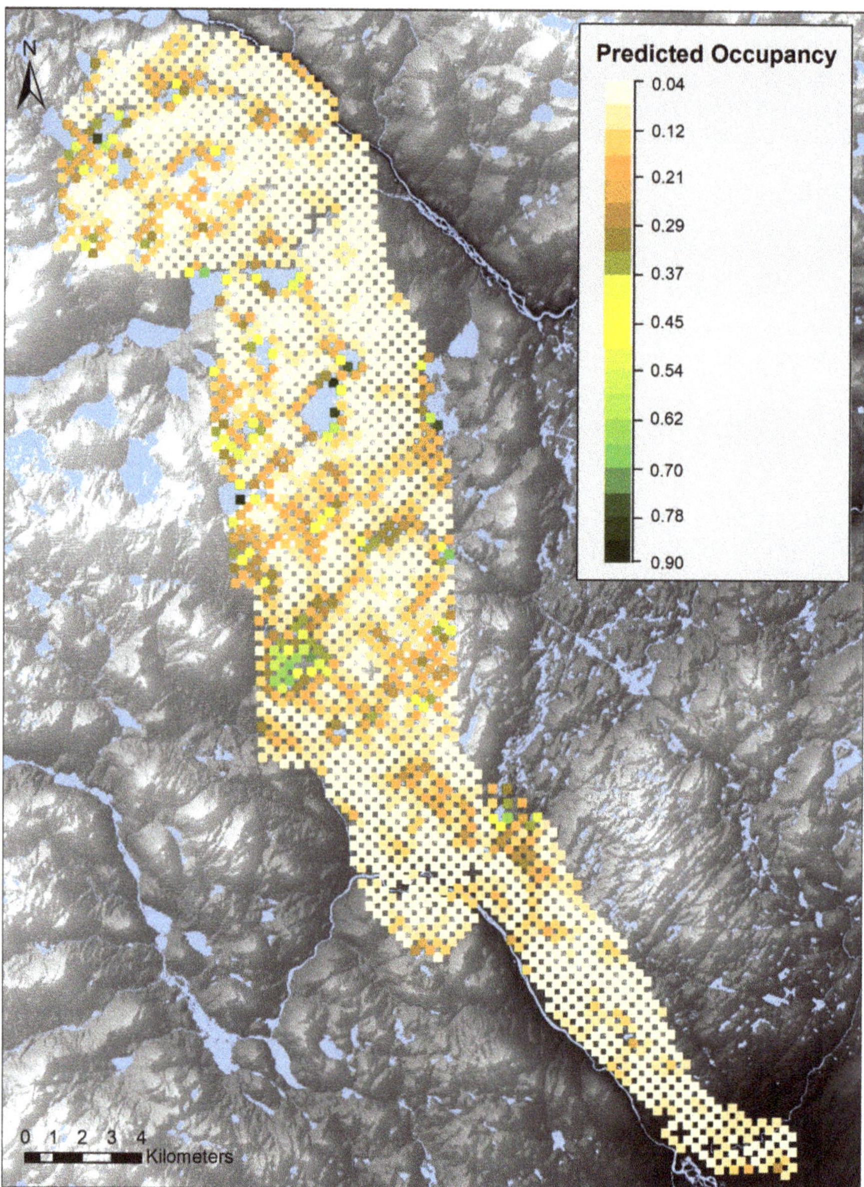

Figure 3. Probability of rusty blackbird occupancy mapped across our study area in Newfoundland, Canada, in 2016 and 2017. Probability of occupancy was estimated from our best model (Table 3) based on landcover data available in the provincial forest resource inventory and publicly available national CanVec stream data.

4. Discussion

Of our 31 *a priori* models, we found that the amount of lakes and ponds, streams, bogs, and rivers best predicted rusty blackbird occurrence. Aquatic invertebrates are an important food source for rusty blackbirds [22,23], and blackbird foraging activities often focus on the edges of waterbodies with abundant shallow water [2]. Thus, the importance of these aquatic habitats in our model is not surprising and is consistent with findings from previous studies of rusty blackbird habitat occupancy [22–24]. However, the absence of conifer forest from our best model and the apparent overall weak influence of either conifer forest or conifer scrub cover on rusty blackbird occupancy initially appears counterintuitive. Powell et al. [24] suggested that blackbird habitat occupancy in Maine, where the forest has a greater component of deciduous cover, is heavily influenced by the presence of patches of suitable conifer habitat in which to nest. In contrast, our study area on Newfoundland is dominated by dense, often single species, conifer stands and consequently rusty blackbirds may not need to actively select for such a ubiquitous feature on the landscape. In particular, wet soils in riparian zones and wetlands on Newfoundland often support dense stands of conifer scrub. To retain landcover features where rusty blackbirds were most likely to occur, we only included forest types that were dominated by either black spruce or balsam fir. Since primarily deciduous stands dominated less than 5% of points, most forested points included conifer cover. Thus, within the spatial scale at which we examined occupancy, rusty blackbirds in Newfoundland appear to select sites primarily based on the availability of appropriate foraging habitat. Additional research such as radio-telemetry studies or nest searching may allow us to better describe the nesting habitat needs of rusty blackbirds in Newfoundland, as well as their space use and breeding ecology. We did not find a strong influence of red squirrel occurrence or year on rusty blackbird occupancy. This is unexpected, given the notable difference in red squirrel detections between years [37]. Red squirrels are known to reproduce later in the season following low cone crop years so that they can take better advantage of seasonal food resources [60], and this influences juvenile survivorship when compared with individuals born earlier in the year [61]. Luepold et al. [22] found that during a year when red squirrel occurrence was higher following a large mast crop, red squirrels were more frequently observed to prey on rusty blackbird nests. It is possible that red squirrels are preying upon rusty blackbird nests, but that despite this, the blackbirds continue to return to habitat where red squirrels are present due to other attractive factors. Red squirrels are prominent members of most boreal forest ecosystems and can directly affect other species through an omnivorous diet and generalist predatory behavior [62,63]. De Santo and Willson [64] found that nest predation was lowest in open wetlands and within forests, compared with both forest and wetland edges and clearcuts. Our observation that red squirrel presence was not strongly related to rusty blackbird occupancy—despite marginally improving our best model in a post-hoc test—may suggest a deviation from the typical vulnerability to nest predation commonly experienced by forest songbirds [63,65,66]. Related research at our study site indicates that squirrels are much more abundant at lower elevations [37] and in these areas they may have caused the local extirpation of breeding populations of another species, the gray-cheeked thrush (McDermott, unpublished data; see also [36,44]). Unlike gray-cheeked thrushes, we found that elevation appeared to have little or no effect on rusty blackbird occupancy. It may be that the impact of squirrels is not as strong for rusty blackbirds as for gray-cheeked thrushes because of a greater capacity to deter predators; for example, rusty blackbirds have been observed mobbing presumed threats (e.g., field crews) near nest sites [53], a trait which is known to drive nest predators away [67]. Alternatively, there may be aspects of their nest site selection that enable them to avoid strong impacts of squirrel predation. Aquatic habitat may also impede squirrel movements during summer and consequently compel squirrels and blackbirds to select different habitats.

McDermott et al. [37] found that squirrel occurrence was inversely related to surface water and ambiguously related to "open" habitats (an amalgamation of bogs, barrens, and other natural openings), which could offer some protection to rusty blackbird nests near those habitat types. Specifically, these open, wet habitats could act as barriers to red squirrel movement. However, the weak relationship between rusty blackbirds and red squirrels could also have resulted from an ecological trap (see Powell et al. [21]), as Luepold et al. [22] found red squirrels to be the most important predator of rusty blackbird nests in New England. More research into rusty blackbird nest success and predator dynamics on Newfoundland may help to clarify the impact of the introduction of red squirrels on this and other boreal bird species.

Consistent with the findings of Powell et al. [24], clearcuts did not appear in our best model set, indicating that this habitat type had little influence on blackbird occupancy. Provincial forestry regulations require that an unharvested buffer strip at least 20-m wide be left along the shorelines of water bodies, and Whitaker and Montevecchi [68] found that, in western Newfoundland, abundance of rusty blackbirds did not differ between these buffer strips and unharvested shorelines. Thus, it may be that current forestry regulations are sufficient to safeguard blackbird habitat needs on Newfoundland. However, because it has been suggested that harvested forests can act as ecological traps for Rusty Blackbirds [21], more research into nesting success of blackbirds at this site could provide a more nuanced perspective into effects of forest management. In addition, the influence of forest management history and temporal patterns at this site is as yet unexplored. This study area provides a unique opportunity to study rusty blackbirds with an abundance of old growth forest. Given that Newfoundland remains a stronghold for rusty blackbirds, it may be worth investigating in more detail the relationship between forest characteristics (i.e., age, structure, diversity) and rusty blackbird abundance.

Our best detectability model included observer as an explanatory variable. Contrary to Powell et al. [24], wind, precipitation, cloud cover, time of day, and ordinal day did not strongly affect detectability, which may reflect differences in the sampling protocol. Variability in detection probability by the observer may reflect false negatives (species was present but was not detected) or false positives (an observer incorrectly identified a species as present; [69–71]). However, false positives are unlikely because similar-looking species (e.g., other blackbirds) are not present on Newfoundland. Intuitively, the frequency of these errors is lower if observers have higher skill levels [70]; all of our observers were skilled in the identification of local species prior to data collection. Given the naïve and predicted blackbird occupancy rates of only 10.7% and 12.2%, respectively, across the study area, differences in detectability between observers may at least in part reflect the potential for some observers to have been assigned survey areas where they were less likely to encounter rusty blackbirds (e.g., due to physical ability or back-country navigation skills). Broadcasting rusty blackbird calls would likely increase detections among all observers because of behavioural responses to conspecific bird calls [24]. In addition, ensuring that all surveyors are of a similar skill level in fitness and backcountry orienteering would improve the consistency of detection rates between observers, since it would ensure that observers cover a similar number of points across a challenging landscape.

The spatial scale at which habitat use is assessed inevitably affects the apparent habitat requirements of a given species [72]; thus, more research on the spatial ecology of rusty blackbirds may lead to improved inferences. Our habitat analysis buffer size reflected the best available home range estimate for rusty blackbirds, although values ranged from 3.8 ha to 172.8 ha [20]. Powell et al. [20] noted that the home range size of colonial rusty blackbirds was significantly larger than that of non-colonial pairs. This likely reflects the potential for birds nesting in loose colonies to share information on the location of short-lived sources of emergent insects, whereas pairs on their own may have more limited sources of food [20]. Therefore, home range size, and a bird's ability to take advantage of available resources may vary drastically based on behavioral factors between individuals. Because we only detected more than two individuals at five out of 209 occupied sites (2.4%) over the two years, it is likely that, as is typical for much of their breeding range, solitary nesting is prevalent at our study area. Concrete evidence of pair versus colonial status is another area where radio-telemetry studies may improve our understanding of rusty blackbird breeding ecology.

Our study is among the first of its kind to model rusty blackbird occupancy using information on habitat from a typical Forest Resource Inventory database that was developed based on high resolution aerial photography, as well as from other publicly available landcover data (e.g., our stream data; but, see Wohner et al. [27]). The fact that we did not undertake field habitat surveys allowed us to efficiently complete a systematic survey of over 1900 point counts across a large area having limited road access. Remote sensing resources such as aerial imagery are considered valuable tools in predicting species distributions and developing population estimates [73,74]. Further, this is the same spatial database that the province uses to plan and monitor industrial forestry and other forms of natural resource management. Consequently, it would be straightforward to use the findings of studies such as this, which are based on information contained in those spatial databases, to predict and map the distribution of rusty blackbirds across the landscape (e.g., Figure 3). This offers the opportunity to easily incorporate consideration of blackbird habitat into conservation, management, and research planning. For example, while the mean predicted occupancy across our study area was 12.2%, the 5% of points having the highest estimates had a mean predicted occupancy of 46.8% (range 32.9%–86.9%); this type of information could be of value in planning research or rapidly identifying high-value habitat during land use planning. Similar forest resource databases are available for many jurisdictions across the North American boreal forest, especially those subject to large scale extractive resource use, so this approach may be applied across much of the species' breeding range to map and protect potential blackbird habitat. Based on our findings, and the findings from other studies on Rusty Blackbird occupancy and habitat use (e.g., [21,22,24,26,27]), key Rusty Blackbird habitat with a high probability of occupancy—such as concentrations of wetlands, waterbodies and watercourses, with nearby dense conifer forest—may be identified from remotely sensed data. Once these areas are identified, prioritization of survey areas and conservation planning may proceed.

5. Conclusions

This is the first quantitative study of rusty blackbirds on the island of Newfoundland, and among the first published studies to use remotely-sensed data to predict their breeding habitat. Given that they may be genetically distinct, this population is important to the overall conservation and recovery of the species. Further, the decline of rusty blackbirds on Newfoundland has been more gradual than in most other areas of the species' breeding range, and they continue to be detected in higher numbers on Newfoundland than elsewhere in their breeding range [7]. Indeed, our naïve occupancy rate was 41.7% higher than that of Powell et al. [24] for a population in northern New England, while our predicted occupancy in the upper 5% of most preferred sites averaged more than four times the overall naïve occupancy rate.

Consistent with past research, our study indicates a strong association for breeding rusty blackbirds with aquatic habitats in the boreal forest, including lakes, ponds, streams, rivers, and bogs; these findings echo the results of Bale et al. [46], and the conservation value of these wetland environments, particularly in the face of climate change. Due to the island's cool, wet maritime climate, boreal landscapes across much of Newfoundland consist of a complex mosaic of bogs and surface water intertwined with coniferous scrub and forests. This appears to offer relatively plentiful habitat for rusty blackbirds, and presents an opportunity to study this declining species in a region where relatively high numbers persist.

Author Contributions: Conceptualization, K.K.E.M., L.L.P., J.P.B.M., D.M.W., I.G.W.; methodology, K.K.E.M., L.L.P., J.P.B.M., D.M.W., I.G.W.; formal analysis, K.K.E.M., L.L.P., D.M.W.; investigation, J.P.B.M., K.K.E.M.; resources, I.W., D.M.W.; data curation, K.K.E.M., J.P.B.M.; writing—original draft preparation, K.K.E.M.; writing—review and editing, K.K.E.M., L.L.P., J.P.B.M., D.M.W., I.G.W; visualization, K.K.E.M., L.L.P., J.P.B.M., D.M.W.; supervision, J.P.B.M., I.G.W., D.M.W., L.L.P.; project administration, J.P.B.M., D.M.W., I.G.W.; funding acquisition, I.G.W., D.M.W.; All authors have read and agreed to the published version of the manuscript.

Funding: This research was funded by the Centre for Forest Science and Innovation (Newfoundland and Labrador Department of Natural Resources), the Natural Sciences and Engineering Research Council of Canada, and Gros Morne National Park of Canada. The Article Processing Charges were funded by Parks Canada.

Acknowledgments: Many thanks to field surveyors Elora Grahame, Brendan Kelly, Noah Korne, Anna Rodgers, Meaghan Tearle, and Benjamin West. Matthew Brooks and Andrew Curtis also provided critical field support, while Jake Burton (Parks Canada) offered geomatics assistance and Dylan Harding helped with data curation. The Newfoundland and Labrador Department of Fisheries and Land Resources, Forestry and Wildlife Branch provided use of a cabin for field surveys and other logistical support. Research was conducted under a scientific research permit from the Newfoundland and Labrador Department of Environment and Conservation, Parks and Natural

Diversity **2020**, *12*, 340

Areas Division, as well as a research permit from the Department of Fisheries and Land Resources, Forestry and Wildlife Branch, and animal care approval (16-16-IW) from the Memorial University of Newfoundland Institutional Animal Care Committee.

Conflicts of Interest: The authors declare no conflict of interest. The funders had no role in the design of the study; in the collection, analyses, or interpretation of data; in the writing of the manuscript, or in the decision to publish the results.

References

1. Committee on the Status of Endangered Wildlife in Canada (COSEWIC). *COSEWIC Assessment and Status Report on the Rusty Blackbird Euphagus carolinus in Canada*; COSEWIC: Ottawa, ON, Canada, 2017; Available online: https://www.canada.ca/en/environment-climate-change/services/species-risk-public-registry/cosewic-assessments-status-reports/rusty-blackbird-2017.html (accessed on 7 May 2020).
2. Greenberg, R.; Droege, S. On the decline of the Rusty Blackbird and the use of ornithological literature to document long-term population trends. *Conserv. Biol.* **1999**, *13*, 553–559. [CrossRef]
3. Greenberg, R.; Matsuoka, S.M. Rangewide ecology of the declining Rusty Blackbird: Mysteries of a species in decline. *Condor* **2010**, *112*, 770–777. [CrossRef]
4. Avery, M.L. Rusty Blackbird (*Euphagus carolinus*), version 1.0. In *Birds of the World*; Poole, A.F., Ed.; Cornell Lab of Ornithology: Ithaca, NY, USA, 2020. [CrossRef]
5. Faaborg, J.; Holmes, R.T.; Anders, A.D.; Bildstein, K.L.; Dugger, K.M.; Gauthreaux, S.A.; Heglund, P.; Hobson, K.A.; Jahn, A.E.; Johnson, D.H. Conserving migratory land birds in the new world: Do we know enough? *Ecol. Appl.* **2010**, *20*, 398–418. [CrossRef]
6. Marra, P.P.; Cohen, E.B.; Loss, S.R.; Rutter, J.E.; Tonra, C.M. A call for full annual cycle research in animal ecology. *Biol. Lett.* **2015**, *11*, 1–4. [CrossRef] [PubMed]
7. Greenberg, R.; Demarest, D.W.; Matsuoka, S.M.; Mettke-Hofmann, C.; Evers, D.; Hamel, P.B.; Luscier, J.; Powell, L.L.; Shaw, D.; Avery, M.L.; et al. Understanding declines in Rusty Blackbirds. In *Boreal Birds of North America: A Hemispheric View of Their Conservation Links and Significance*; Studies in Avian Biology (no. 41); Wells, J.V., Ed.; University of California Press: Berkeley, CA, USA, 2011; pp. 107–126.
8. Edmonds, S.T.; Evers, D.C.; Cristol, D.A.; Mettke-Hofmann, C.; Powell, L.L.; McGann, A.J.; Armiger, J.W.; Lane, O.P.; Tessler, D.F.; Newell, P.; et al. Geographic and seasonal variation in mercury exposure of the declining Rusty Blackbird. *Condor* **2010**, *112*, 789–799. [CrossRef]
9. Hobson, K.A.; Bayne, E.M.; van Wilgenburg, S.L. Large-scale conversion of forest to agriculture in the Boreal Plains of Saskatchewan. *Conserv. Biol.* **2002**, *16*, 1530–1541. [CrossRef]
10. Morissette, J.L.; Bayne, E.M.; Kardynal, K.J.; Hobson, K.A. Regional variation in responses of wetland-associated bird communities to conversion of boreal forest to agriculture. *Avian Conserv. Ecol.* **2019**, *14*, 12. [CrossRef]
11. McClure, C.J.W.; Rolek, B.W.; McDonald, K.; Hill, G.E. Climate change and the decline of a once common bird. *Ecol. Evol.* **2012**, *2*, 370–378. [CrossRef]
12. Stralberg, D.; Matsuoka, S.M.; Hamann, A.; Bayne, E.M.; Solymos, P.; Schmiegelow, F.K.A.; Wang, X.; Cumming, S.G.; Song, S.J. Projecting boreal bird responses to climate change: The signal exceeds the noise. *Ecol. Appl.* **2015**, *25*, 1112–1128. [CrossRef]
13. Holmes, R.T.; Marra, P.P.; Sherry, T.W. Habitat specific demography of breeding Black-throated Blue Warblers (*Dendroica caerulescens*): Implications for population dynamics. *J. Anim. Ecol.* **1996**, *65*, 183–195. [CrossRef]
14. Holt, C.A.; Fuller, R.J.; Dolman, P.M. Deer reduce habitat quality for a woodland songbird: Evidence from settlement patterns, demographic parameters, and body condition. *Auk* **2013**, *130*, 13–20. [CrossRef]
15. Matsuoka, S.M.; Handel, C.M.; Ruthrauff, D.R. Densities of breeding birds and changes in vegetation in an Alaskan boreal forest following a massive disturbance by spruce beetles. *Can. J. Zool.* **2001**, *79*, 1678–1690. [CrossRef]
16. Nappi, A.; Drapeau, P.; Saint-Germain, M.; Angers, V.A. Effect of fire severity on long-term occupancy of burned boreal conifer forests by saproxylic insects and wood-foraging birds. *Int. J. Wildland Fire* **2009**, *19*, 500–511. [CrossRef]
17. Hannon, S.J.; Schmiegelow, F.K.A. Corridors may not improve the conservation value of small reserves for most boreal birds. *Ecol. Appl.* **2002**, *12*, 891–899. [CrossRef]

18. Potvin, F.; Bertrand, N. Leaving forest strips in large clearcut landscapes of boreal forest: A management scenario suitable for wildlife? *For. Chron.* **2004**, *80*, 44–53. [CrossRef]

19. Ellison, W.G. *The Status and Habitat of the Rusty Blackbird in Caledonia and Essex Counties*; Vermont Fish and Wildlife Department: Woodstock, VT, USA, 1990.

20. Powell, L.L.; Hodgman, T.P.; Glanz, W.E. Home ranges of Rusty Blackbirds breeding in wetlands: How much would buffers from timber harvest protect habitat? *Condor* **2010**, *112*, 834–840. [CrossRef]

21. Powell, L.L.; Hodgman, T.P.; Glanz, W.E.; Osenton, J.D.; Fisher, C.M. Nest-site selection and nest survival of the Rusty Blackbird: Does timber management adjacent to wetlands create ecological traps? *Condor* **2010**, *112*, 800–809. [CrossRef]

22. Luepold, S.H.B.; Hodgman, T.P.; McNulty, S.A.; Cohen, J.; Foss, C.R. Habitat selection, nest survival, and nest predators of Rusty Blackbirds in northern New England, USA. *Condor* **2015**, *117*, 609–623. [CrossRef]

23. Matsuoka, S.M.; Shaw, D.; Johnson, J.A. Estimating the abundance of nesting Rusty Blackbirds in relation to the wetland habitats in Alaska. *Condor* **2010**, *112*, 825–833. [CrossRef]

24. Powell, L.L.; Hodgman, T.P.; Fiske, I.J.; Glanz, W.E. Habitat occupancy of Rusty Blackbirds (*Euphagus carolinus*) breeding in northern New England, USA. *Condor* **2014**, *116*, 122–133. [CrossRef]

25. Matsuoka, S.M.; Shaw, D.; Sinclair, P.H.; Johnson, J.A.; Corcoran, R.M.; Dau, N.C.; Meyers, P.M.; Rojek, N.A. Nesting ecology of the Rusty Blackbird in Alaska and Canada. *Condor* **2010**, *112*, 810–824. [CrossRef]

26. Buckley, S.H. Rusty Blackbirds in Northeastern Industrial Forests: A Multi-scale Study of Nest Habitat Selection and Nest Survival. Master's Thesis, SUNY College of Environmental Sciences and Forestry, Syracuse, NY, USA, April 2013.

27. Wohner, P.J.; Foss, C.R.; Cooper, R.J. Rusty Blackbird habitat selection and survivorship during nesting and post-fledging. *Diversity* **2020**, *12*, 221. [CrossRef]

28. Morosinotto, C.; Thomson, R.L.; Korpimäki, E. Habitat selection as an antipredator behaviour in a multi-predator landscape: All enemies are not equal. *J. Anim. Ecol.* **2010**, *79*, 327–333. [CrossRef] [PubMed]

29. Siepielski, A.M. A possible role for red squirrels in structuring breeding bird communities in lodgepole pine forests. *Condor* **2006**, *108*, 232–238. [CrossRef]

30. McFarland, K.P.; Rimmer, C.C.; Frey, S.J.K.; Faccio, S.D.; Collins, B.B. *Demography, Ecology and Conservation of Bicknell's Thrush in Vermont, with a Special Focus on the Northeast Highlands*; Technical Report 08-03; Vermont Center for Ecostudies: Norwich, VT, USA, 2008; Available online: https://www.researchgate.net/publication/228412299 (accessed on 7 May 2020).

31. Haché, S.; Bayne, E.M.; Villard, M.-A. Postharvest regeneration, sciurid abundance, and postfledging survival and movements in an Ovenbird population. *Condor* **2014**, *116*, 102–112. [CrossRef]

32. Dodds, D. Terrestrial mammals. In *Biogeography and Ecology of the Island of Newfoundland*; South, R., Ed.; Springer: The Hague, The Netherlands, 1983; pp. 163–206.

33. Lewis, K.P. Processes Underlying Nest Predation by Introduced Red Squirrels (*Tamiasciurus hudsonicus*) in the Boreal Forest of Newfoundland. Ph.D. Thesis, Memorial University of Newfoundland, St. John's, NL, Canada, May 2004.

34. Whitaker, D. The colonisation of Newfoundland by red squirrels (*Tamiasciurus hudsonicus*); chronology, environmental effects and future needs. *Osprey* **2015**, *46*, 23–29.

35. Committee on the Status of Endangered Wildlife in Canada (COSEWIC). *COSEWIC Assessment and Status Report on the Red Crossbill Percna Subspecies Loxia Curvirostra Percna in Canada*; COSEWIC: Ottawa, ON, Canada, 2016; Available online: https://www.canada.ca/en/environment-climate-change/services/species-risk-public-registry/cosewic-assessments-status-reports/red-crossbill-2016.html (accessed on 7 May 2020).

36. Whitaker, D.M.; Taylor, P.D.; Warkentin, I.G. Gray-cheeked Thrush (*Catharus minimus minimus*) distribution and habitat use in a montane forest landscape of western Newfoundland, Canada. *Avian Conserv. Ecol.* **2015**, *10*, 4. [CrossRef]

37. McDermott, J.P.B.; Whitaker, D.M.; Warkentin, I.G. Constraints on range expansion of introduced red squirrels (*Tamiasciurus hudsonicus*) in an island ecosystem. *Can. J. For. Res.* **2020**. [CrossRef]

38. Benkman, C.W. On the evolution and ecology of island populations of crossbills. *Evolution* **1989**, *43*, 1324–1330. [CrossRef]

39. Robertson, B.A.; Hutto, R.L. Is selectively harvested forest an ecological trap for olive-sided flycatchers? *Condor* **2007**, *109*, 109–121. [CrossRef]

40. Vitz, A.C.; Rodewald, A.D. Behavioral and demographic consequences of access to early-successional habitat in juvenile Ovenbirds (*Seiurus aurocapilla*): An experimental approach. *Auk* **2013**, *130*, 21–29. [CrossRef]

41. Charchuk, C.; Bayne, E.M. Avian community response to understory protection harvesting in the boreal forest of Alberta, Canada. *For. Ecol. Manag.* **2018**, *407*, 9–15. [CrossRef]

42. Peters, H.S.; Burleigh, T.D. *The Birds of Newfoundland*; The Riverside Press: Cambridge, MA, USA, 1951.

43. Montevecchi, W.A.; Tuck, L.M. *Newfoundland Birds: Exploitation, Study, Conservation*; Nuttall Ornithological Club: Cambridge, MA, USA, 1987; ISBN 9997639995.

44. FitzGerald, A.M.; Whitaker, D.M.; Ralston, J.; Kirchman, J.J.; Warkentin, I.G. Taxonomy and distribution of the imperilled Newfoundland Gray-cheeked Thrush, *Catharus minimus minimus*. *Avian Conserv. Ecol.* **2017**, *12*, 10. [CrossRef]

45. Burleigh, T.D.; Peters, H.S. Geographic variation in Newfoundland birds. *Proc. Biol. Soc. Wash.* **1948**, *61*, 111–124.

46. Bale, S.; Beazley, K.F.; Westwood, A.; Bush, P. The benefits of using topographic features to predict climate-resilient habitat for migratory forest landbirds: An example for the Rusty Blackbird, Olive-sided Flycatcher, and Canada Warbler. *Condor* **2020**, *122*, 1–19. [CrossRef]

47. Thompson, I.D.; Larson, D.J.; Montevecchi, W.A. Characterization of old "wet boreal" forests, with an example from balsam fir forests of western Newfoundland. *Environ. Rev.* **2003**, *11*, S23–S46. [CrossRef]

48. McCarthy, J.W.; Weetman, G. Age and size structure of gap-dynamic, old-growth boreal forest stands in Newfoundland. *Silva Fenn.* **2006**, *40*, 209–230. [CrossRef]

49. Viglas, J.N.; Brown, C.D.; Johnstone, J.F. Age and size effects on seed productivity in northern black spruce. *Can. J. For. Res.* **2013**, *43*, 534–543. [CrossRef]

50. Arsenault, A.; LeBlanc, R.; Earle, E.; Brooks, D.; Clarke, B.; Lavigne, D.; Royer, L. Unravelling the past to manage Newfoundland's forest for the future. *Forest. Chron.* **2016**, *92*, 487–502. [CrossRef]

51. Environment Canada. North American Breeding Bird Survey Instructions and Safety Guidelines. 2017. Available online: https://www.canada.ca/content/dam/eccc/migration/main/reom-mbs/5ee0adba-a60b-4142-9add-644f35e5935e/bbs_instructions_formatted_en.pdf (accessed on 22 July 2020).

52. Ralph, C.J.; Sauer, J.R.; Droege, S. *Monitoring Bird Populations by Point Count*; United States Department of Agriculture: Berkeley, CA, USA, 1997. Available online: https://www.fs.fed.us/psw/publications/documents/psw_gtr149/psw_gtr149.pdf (accessed on 16 June 2020).

53. Powell, L.L.; Hodgman, T.P.; Glanz, W.E.; Osenton, J.D.; Ellis, D.E. A loose colony of Rusty Blackbirds nesting in Northern Maine. *Northeast. Nat.* **2010**, *17*, 639–646. [CrossRef]

54. ESRI. *ArcGIS Desktop*; (10.5.1); Computer Software; Environmental Systems Research Institute: Redlands, CA, USA, 2017.

55. Dormann, C.F.; Elith, J.; Bacher, S.; Buchmann, C.; Carré, G.C.G.; Garcia Márquez, J.R.; Gruber, B.; Lafourcade, B.; Leitao, P.J.; Münkemüller, T.; et al. Collinearity: A review of methods to deal with it and a simulation study evaluating their performance. *Ecography* **2013**, *36*, 27–46. [CrossRef]

56. Fiske, I.; Chandler, R. Unmarked: An R package for fitting hierarchical models of wildlife occurrence and abundance. *J. Stat. Softw.* **2011**, *43*, 1–23. [CrossRef]

57. R Core Team. *R: A Language and Environment for Statistical Computing*, version 3.5.1; R Foundation for Statistical Computing: Vienna, Austria, 2013.

58. Mackenzie, D.I.; Nichols, J.D.; Lachman, G.B.; Droege, S.; Royle, J.A.; Langtimm, C.A. Estimating site occupancy rates when detection probabilities are less than one. *Ecology* **2002**, *83*, 2248–2255. [CrossRef]

59. Burnham, K.P.; Anderson, D.R. *Model Selection and Multimodel Inference*, 2nd ed.; Springer: New York, NY, USA, 2002.

60. Fletcher, Q.E.; Landry-Cuerrier, M.; Boutin, S.; McAdam, A.G.; Speakman, J.R.; Humphries, M.M. Reproductive timing and reliance on hoarded capital resources by lactating red squirrels. *Oecologia* **2013**, *173*, 1203–1215. [CrossRef] [PubMed]

61. Williams, C.T.; Lane, J.E.; Humphries, M.M.; McAdam, A.G.; Boutin, S. Reproductive phenology of a food-hoarding mast-seed consumer: Resource- and density-dependent benefits of early breeding in red squirrels. *Oecologia* **2013**, *174*, 777–788. [CrossRef]

62. Martin, T.E. Processes organizing open-nesting bird assemblages: Competition or nest predation? *Evol. Ecol.* **1988**, *2*, 37–50. [CrossRef]

63. Bayne, E.M.; Hobson, K.A. Effects of red squirrel (*Tamiasciurus hudsonicus*) removal on survival of artificial songbird nests in boreal forest fragments. *Am. Midl. Nat.* **2002**, *147*, 72–79. [CrossRef]

64. De Santo, T.L.; Willson, M.F. Predator abundances and predation of artificial nests in natural and anthropogenic coniferous forest edges in southeast Alaska. *J. Field Ornithol.* **2001**, *72*, 136–149. [CrossRef]

65. Willson, M.F.; de Santo, T.L.; Sieving, K.E. Red squirrels and predation risk to bird nests in northern forests. *Can. J. Zool.* **2003**, *81*, 1202–1208. [CrossRef]

66. Sherry, T.W.; Wilson, S.; Hunter, S.; Holmes, R.T. Impacts of nest predators and weather on reproductive success and population limitation in a long-distance migratory songbird. *J. Avian Biol.* **2015**, *46*, 559–569. [CrossRef]

67. Krama, T.; Bērziņš, A.; Rytkönen, S.; Rantala, M.J.; Wheatcroft, D.; Krams, I. Linking anti-predator behaviour and habitat quality: Group effect in nest defence of a passerine bird. *Acta Ethol.* **2012**, *15*, 127–134. [CrossRef]

68. Whitaker, D.M.; Montevecchi, W.A. Breeding bird assemblages inhabiting riparian buffer strips in Newfoundland, Canada. *J. Wildl. Manag.* **1999**, *63*, 167–179. [CrossRef]

69. Bart, J. Causes of recording errors in singing bird surveys. *Wilson Bull.* **1985**, *97*, 161–172.

70. Farmer, R.G.; Leonard, M.L.; Horn, A.G. Observer effects and avian-call-count survey quality: Rare-species biases overconfidence. *Auk* **2012**, *129*, 76–86. [CrossRef]

71. Royle, J.A.; Link, W.A. Generalized site occupancy models allowing for false positive and false negative errors. *Ecology* **2006**, *87*, 835–841. [CrossRef]

72. Weins, J.A.; Rotenberry, J.T.; van Horne, B. Habitat occupancy patterns of North American shrubsteppe birds: The effects of spatial scale. *Oikos* **1987**, *48*, 132–147. [CrossRef]

73. Gottschalk, T.K.; Huettmann, F.; Ehlers, M. Thirty years of analysing and modelling avian habitat relationships using satellite imagery data: A review. *Int. J. Remote Sens.* **2007**, *26*, 2631–2656. [CrossRef]

74. Shealer, D.A.; Alexander, M.J. Use of aerial imagery to assess habitat suitability and predict site occupancy for a declining wetland-dependent bird. *Wetl. Ecol. Manag.* **2013**, *21*, 289–296. [CrossRef]

Article

Rusty Blackbird (*Euphagus carolinus*) Foraging Habitat and Prey Availability in New England: Implications for Conservation of a Declining Boreal Bird Species

Amanda Pachomski [1],*, Stacy McNulty [2],*, Carol Foss [3], Jonathan Cohen [1] and Shannon Farrell [1]

[1] Department of Environmental and Forest Biology, State University of New York College of Environmental Science and Forestry, 1 Forestry Drive, Syracuse, NY 13210, USA; jcohen14@esf.edu (J.C.); sfarrell@esf.edu (S.F.)

[2] Adirondack Ecological Center, State University of New York College of Environmental Science and Forestry, 6312 State Route 28N, Newcomb, NY 12852, USA

[3] New Hampshire Audubon Society, 84 Silk Farm Road, Concord, NH 03301, USA; cfoss@nhaudubon.org

* Correspondence: amandapachomski@gmail.com (A.P.); smcnulty@esf.edu (S.M.)

Citation: Pachomski, A.; McNulty, S.; Foss, C.; Cohen, J.; Farrell, S. Rusty Blackbird (*Euphagus carolinus*) Foraging Habitat and Prey Availability in New England: Implications for Conservation of a Declining Boreal Bird Species. *Diversity* **2021**, *13*, 99. https://doi.org/10.3390/d13020099

Academic Editor: Michael Wink

Received: 31 December 2020
Accepted: 18 February 2021
Published: 23 February 2021

Publisher's Note: MDPI stays neutral with regard to jurisdictional claims in published maps and institutional affiliations.

Abstract: The Rusty Blackbird (*Euphagus carolinus*) is an imperiled migratory songbird that breeds in and near the boreal wetlands of North America. Our objective was to investigate factors associated with Rusty Blackbird wetland use, including aquatic invertebrate prey and landscape features, to better understand the birds' habitat use. Using single-season occupancy modeling, we assessed breeding Rusty Blackbird use of both active and inactive beaver-influenced wetlands in New Hampshire and Maine, USA. We conducted timed, unlimited-radius point counts of Rusty Blackbirds at 60 sites from May to July 2014. Following each point count, we sampled aquatic invertebrates and surveyed habitat characteristics including percent mud cover, puddle presence/absence, and current beaver activity. We calculated wetland size using aerial imagery and calculated percent conifer cover within a 500 m buffer of each site using the National Land Cover Database 2011. Percent mud cover and invertebrate abundance best predicted Rusty Blackbird use of wetlands. Rusty Blackbirds were more likely to be found in sites with lower percent mud cover and higher aquatic invertebrate abundance. Sites with Rusty Blackbird detections had significantly higher abundances of known or likely prey items in the orders Amphipoda, Coleoptera, Diptera, Odonata, and Trichoptera. The probability of Rusty Blackbird detection was 0.589 ± 0.06 SE. This study provides new information that will inform habitat conservation for this imperiled species in a beaver-influenced landscape.

Keywords: Rusty Blackbird; *Euphagus carolinus*; boreal wetlands; aquatic macroinvertebrates; foraging ecology; occupancy modeling

1. Introduction

The Rusty Blackbird (*Euphagus carolinus*) is a migratory songbird that breeds in and near wetlands of the boreal forests of Canada and Alaska as well as in the northern regions of New York and the Acadian Forest (New England and the Canadian Maritime Provinces). The Rusty Blackbird is representative of global boreal avian species declines and has experienced the worst population loss of thirteen boreal-breeding species [1,2]. Although the Rusty Blackbird was once common, the species has declined by an estimated 90% since the 1960s [3]. The species was estimated to have declined by 5.1% per year from over 13 million birds in 1965–1966 to roughly 2 million birds in 2002–03 based on modeling from standardized winter counts [1]. Furthermore, the southeastern limits of the bird's breeding range appear to have retreated northward coincident with the population decline [4]. The US Fish and Wildlife Service has listed the Rusty Blackbird as a Focal Species of Birds of Management Concern [5]; the IUCN Red List considers the species to be Vulnerable [6]. The cause of the Rusty Blackbird's decline is not fully understood; climate change [4], mercury contamination [7], hematozoa infections [8], timber harvesting on breeding grounds [9], and winter habitat loss [10] have been suggested as possible factors.

As with all species, suitable foraging and nesting habitat are key components for Rusty Blackbird persistence. In New England, Rusty Blackbirds select nest sites with minimal canopy cover, high basal area of young conifers (<1.5 m height) [9,11], and patches of forest adjacent to open areas, including wetlands [11]. They build cup nests in live trees, usually red spruce (*Picea rubens*), black spruce (*Picea mariana*), or balsam fir (*Abies balsamea*), surrounded by other young conifers, and occasionally in speckled alder (*Alnus incana*) swamps, in snags, or in isolated conifers in open areas [11]. Rusty Blackbird habitat can be created or improved by forest management or natural disturbance, especially by the American beaver (*Castor canadensis*). The beaver, an ecosystem engineer, creates impoundments of water by damming streams [12], forming wetlands utilized by a diversity of wildlife species. Because deciduous trees and shrubs are a preferred food of the beaver [13], they selectively harvest several woody species in proximity to their impoundments, thereby increasing the percent cover of conifers [14], which is desirable for Rusty Blackbird nesting. Beavers have a long-lasting impact on the landscape; by digging channels and creating dams, beavers increase both the depth and area of wetlands [15]. Individual beaver ponds may be active for one to many years at a time as the animals move to find new sources of food, resulting in a matrix of different-aged ponds, meadows, and streams in a wetland complex [16]. Beavers increase the diversity of and shift the macroinvertebrate assemblage [17,18] as well as increase macroinvertebrate abundance [18] within impounded wetlands and streams. Thus, beavers may create an ideal habitat for Rusty Blackbirds, with flooded, macroinvertebrate-rich wetlands for foraging and clumps of nearby conifers for nesting.

It is critical to monitor Rusty Blackbird populations and habitat use in order to make and evaluate the result of management decisions. Previous research within the Rusty Blackbird's breeding range has focused on demography, nesting ecology, and possible causes of decline. Information about the species' diet and foraging site preferences are scant. Rusty Blackbirds are more insectivorous than other Icterids, based on their skull and bill anatomy [19] and analysis of stomach contents [19–22]. They forage aerially, from perches, and while walking along the water's edge. A breeding Rusty Blackbird diet consists mostly of aquatic macroinvertebrates, such as beetle adults and larvae [22,23], Odonate (dragonfly and damselfly) larvae [24], Trichoptera (caddisfly) larvae and emergent adults [25], and Tipulid (crane fly) larvae [25], but they also hunt aerial prey such as mosquitoes [26]. Breeding Rusty Blackbirds also forage for a variety of terrestrial and volant invertebrates, including snails [27], grasshoppers [20], caterpillars [20], spiders [20,27,28], adult dragonflies [25,29], adult mayflies [29], ants, centipedes, and crustaceans [22]. Furthermore, although Rusty Blackbirds are mostly insectivorous during the breeding season [30], the species has been known to eat some vertebrates such as small fish [27,28] and salamanders [27]. In addition to consumption of various prey items during the breeding season, Rusty Blackbirds exhibit diet plasticity by switching to a more generalized diet of seeds, fruit, acorns, grains, and insects during autumn and winter [22,23]. Overall, summer feeding observations and stomach specimens are very limited in number, and little is known about their foraging habitat requirements. Understanding wetland prey availability during the breeding season in key patches will enable scientists and land managers to identify high-quality foraging sites and may eventually suggest mechanisms behind the Rusty Blackbird decline and potential for recovery.

Rusty Blackbirds are rare and difficult to detect, especially within their remote and hard-to-access breeding grounds; thus, traditional, short duration (five or ten minutes) avian point-counts are not sufficient for accurately detecting breeding Rusty Blackbirds [3]. We used a single-season site occupancy modeling approach [31] to account for imperfect detection and model Rusty Blackbird use of 60 wetland sites as a function of habitat covariates, with a focus on foraging habitat and food availability. We chose multiple a priori habitat covariates that we expected to be biologically important for Rusty Blackbirds based on previous studies and our own experience. We hypothesized that the probability of Rusty Blackbird wetland use: (1) increases with current beaver activity, (2) increases with the presence of puddles, and (3) increases with increasing conifer cover. We hypothesized that the probability of detecting a present Rusty Blackbird (1) is not affected by the time of day during daylight hours, (2) decreases with increasing wind speed, (3) is highest during the chick-rearing period, and (4) decreases with increasing wetland size. This study is the first to assess Rusty Blackbird foraging habitat use in New England and the first to include prey availability as a covariate in Rusty Blackbird occupancy modeling.

2. Materials and Methods

2.1. Study Area

We surveyed breeding Rusty Blackbird use of both active and inactive beaver- influenced wetlands in Coos County, New Hampshire, and Oxford County, Maine (Figure 1). Sites were located either on federal land at Umbagog National Wildlife Refuge (44.832344, −71.075496) or were privately owned and managed by Wagner Forest Management, Ltd. The study area has a mean precipitation of 102.9 cm per year, a mean annual high temperature of 11.67 °C, and a mean low temperature of −2.22 °C (measured in Colebrook, NH) [32]. The mean elevation of surveyed wetlands was 473.3 m (range = 110 m to 780 m). Beavers have been modifying wetland hydrology and upland vegetation in the region for decades; survey sites were categorized as either active (impounded/modified in the past year and hosting a resident beaver colony) or inactive (previously impounded, but not currently occupied by beavers) wetlands. The forests in this remote area of New England are extensively managed, with active logging operations occurring near most of our study sites.

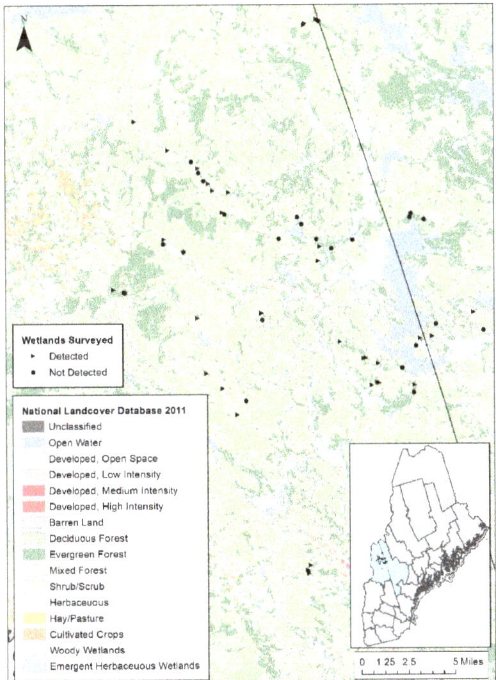

Figure 1. Land cover classes of the study area and digitized polygons of wetlands where Rusty Blackbirds were detected (black triangles) or were not detected (black circles) in northern New England, USA, with 500 m buffer circles showing habitat in 2014.

2.2. Methods

To evaluate Rusty Blackbirds' use of wetlands, we conducted timed, 30 min long, unlimited-radius point counts at 60 sites in 2014 within a 25 km radius of the town center of Errol, NH. We selected count locations from a pool of 263 discrete (e.g., above a dam) beaver-influenced impoundments within 500 m of a road, which we identified using satellite imagery and expert knowledge of the study area. We used ArcMap to randomly select 39 wetland sites and then systemically selected 21 additional, nearby wetlands to maximize our sample size. Bird surveys were conducted from a fixed, marked point located at the southern edge of each wetland. Some sites were located relatively close together (100–300 m apart), so to avoid double counting, pairs of observers concurrently surveyed adjacent sites, cross-checked the timing of field observations field post-survey, and excluded possible double counts of birds. Due to time and logistical constraints, we surveyed sites in pairs based on their spatial proximity; thus, sites were not randomly visited.

We surveyed each point three times, once during each of three two-week intervals (14 May to 27 May, 28 May to 10 June, and 11 June to 24 June 2014). Due to site access restrictions, two of the 60 sites were only surveyed twice. We timed the survey intervals to align with Rusty Blackbird incubation, nestling rearing, and fledging stages of the breeding season because we suspected that detectability would vary based on the birds' nesting behavior.

We conducted surveys between 8:00 and 18:00. We chose not to limit our surveys to early morning hours in order to maximize our sample size and because Rusty Blackbirds are known to sing throughout the day [33]. Through pilot surveys, we found that a point count duration of 30 min was the most effective yet efficient count duration for detecting Rusty Blackbirds in our study area [33]. During each point count, two independent observers recorded visual and/or auditory detections of Rusty Blackbirds and time to first detection (if detected) without the use of any recorded playback. We also recorded wind speed, temperature, precipitation, and time of day (Table 1).

Table 1. Site and field survey covariates used to model detectability and site occupancy of Rusty Blackbirds in northern New England in 2014; the data collection method was by remote sensing (GIS) or at the site (Field).

Covariate	Description	Method
Beaver	Binary measure of observed beaver activity	Field
Day	Ordinal survey date	Field
Elevation	Site elevation in meters	GIS
Water.depth	Average depth (centimeters) of open water near pond edge	Field
Invert.abundance	Average number of invertebrate individuals observed in three samples	Field
Invert.richness	Total number of invertebrate families observed in three samples	Field
Min	Survey start time converted to minute of day	Field
Mud	Visual estimate of percent exposed mud within a wetland	Field
Open.water	Visual estimate of percent open water within a wetland	Field
Pct.conifer	Percent conifer cover within a 500-m buffer of wetland using NLCD 2011	GIS
Puddles	Binary measure of puddles observed 0, 1, 2, or 3 times out of three surveys	Field
Precip	Binary measure of presence of precipitation during survey	Field
Size	Wetland size measured in meters squared	GIS
Temp	Measure of temperature (Celsius) at the start of the survey	Field
Visit	Survey period 1, 2, or 3	Field
Wind	Wind speed (km per hour) measured using an anemometer held at 4.5 m height at the start of the survey	Field
Yng.conifer	Binary measure of presence of dense regenerating spruce and/or fir trees < 1.5 m tall	Field

We collected field data on both vegetation and land cover variables, including percent spruce and balsam fir cover around the wetland, percent cover of open water, percent cover of mud, depth of open water, as well as evidence of recent beaver activity (Table 1). We used the perimeter-based cover estimate method [34] to measure the percent exposed mud and percent open water within the wetland during each survey occasion and then averaged the habitat data across the three occasions.

We collected aquatic invertebrate samples at the point count site (southern edge of each wetland's standing pool of water) with ten sweeps of a D-frame dip net, probing along the water's edge. Macroinvertebrate samples were stored in 70% ethanol and identified to family when possible, otherwise to order or subclass [35]. We used the macroinvertebrate count as a proxy for abundance of Rusty Blackbird prey within a wetland. We averaged the counts for each of the sites' three invertebrate samples and included this abundance as a site covariate in our wetland use models. We also included the total invertebrate family richness for all of each site's samples. For taxa that we were only able to be identified to subclass or order, such as leeches, we assumed that one family was observed for each subclass or order.

We used ArcMap to calculate wetland size and quantify land cover type as habitat characteristics that might drive Rusty Blackbird foraging site selection. Because Rusty Blackbirds forage within multiple wetland types, we chose not to use a wetland classification system and instead delineated all wetland features within a site as a single polygon, using Google Earth satellite imagery (dated 18 September 2013) and field experience as a guide. We then calculated the area of each wetland polygon and used the National Land Cover Database 2011 (NLCD) [36] to calculate the percent conifer cover within a 500 m radius of each wetland, which is the approximate size of a typical Rusty Blackbird breeding home range [11]. We recorded the elevation of each wetland survey point in the field using a GPS unit.

2.3. Analysis

We used the protocol developed by MacKenzie et al. [31] to model Rusty Blackbird use of wetlands using single-species occupancy modeling based on our detection histories and field and geospatial data (Table 1). Occupancy probability (Ψ) and detection probability (p) were modeled as linear functions of covariates using the logit link to constrain predicted values between 0 and 1. These models assume a closed population between surveys, and where that assumption is violated, Ψ is interpreted as habitat use rather than occupancy. Because Rusty Blackbirds may forage among multiple wetlands within a large area, defining a site as a single wetland, as we did, may violate the assumption of independence of observations. Furthermore, concurrent Rusty Blackbird productivity research revealed that Rusty Blackbirds nested near many of our study sites. However, breeding Rusty Blackbirds in New England have been found to nest up to 400 m away from wetlands, and fledglings can move over 1 km away from their nests within a few weeks of fledging [37], so we were not able to assume that site occupancy was constant throughout the study period. Thus, we considered sites with at least one positive Rusty Blackbird detection to be "used" rather than "occupied" [38].

We performed this analysis using Package unmarked [39] in Program R [40]. Our candidate set of models included biologically plausible variables known or thought to affect Rusty Blackbird habitat suitability. We first modeled survey-specific covariates affecting p (date, precipitation, temperature, time, visit, wetland size, and wind) while modeling Ψ as constant. Then, we chose the model with the lowest Akaike Information Criterion (AIC) score as the best-fit detectability model and used its survey covariate(s) in our base Ψ model, following the approach of Powell et al., 2014 [41]. Next, we created a candidate set of models with one or more site covariates (beaver, invertebrate abundance, invertebrate richness, open water, mud, puddles, water depth, percent conifer, young conifer, elevation, and wetland size). We avoided including significantly correlated ($p < 0.05$) covariates (calculated using Spearman Rho and Pearson Chi-Square tests for continuous and categorical covariates, respectively) within the same wetland use models. We used Package AICcmodavg [42] to estimate c-hat, the overdispersion parameter, to adjust for overdispersion as needed, and to assess model fit. We used the MacKenzie and Bailey Goodness-of-fit Test [43] to test the fit of our global wetland use model, which contained all covariates included in our candidate set of models, with 1000 bootstraps.

To assess whether Rusty Blackbirds select foraging sites based on the abundance of specific aquatic invertebrate prey types, we used the two-sample Poisson rate test to test for a difference between the maximum invertebrate abundance per order at sites with and without Rusty Blackbird detections. We included underrepresented invertebrate groups in abundance totals in order to reflect total prey availability at each site.

3. Results

We detected Rusty Blackbirds during 66 of 178 surveys. Our base detectability model (without covariates) yielded a detection probability (p) of 0.589 ± 0.06 SE (95% CI: -0.10, 0.82). Detection probability was best predicted by visit (survey period), which was the only covariate in the top model (Table 2). The probability of detection was highest in visit 2 and lowest in visit 1. Back-transformed parameter estimates on the probability scale for visit yielded $p = 0.416 \pm 0.09$ SE for visit 1, $p = 0.765 \pm 0.08$ SE for visit 2, and $p = 0.742 \pm 0.09$ SE for visit 3.

Table 2. Model selection for detectability of Rusty Blackbirds in northern New England in 2014.

Model	K [a]	AIC [b]	ΔAIC [c]	wi [d]	−2 Log-Likelihood
p(visit) Ψ(.)	4	214.38	0.00	0.513	206.4
p(size) Ψ(.)	3	218.18	3.80	0.077	212.2
p(precip) Ψ(.)	3	219.19	4.81	0.046	213.2
p(temp) Ψ(.)	3	219.26	4.88	0.045	213.2
p(time) Ψ(.)	3	219.34	4.96	0.043	213.3
p(day) Ψ(.)	3	219.39	5.01	0.042	213.4
p(wind) Ψ(.)	3	219.50	5.12	0.040	213.5
p(day+precip) Ψ(.)	4	220.87	6.49	0.020	212.9
p(time+precip) Ψ(.)	4	220.96	6.58	0.019	213.0
p(day+time) Ψ(.)	4	220.99	6.61	0.019	213.0
p(temp+wind) Ψ(.)	4	221.05	6.67	0.018	213.1
p(time+temp) Ψ(.)	4	221.07	6.69	0.018	213.1
p(day+temp) Ψ(.)	4	221.08	6.70	0.018	213.1
p(time+wind) Ψ(.)	4	221.19	6.81	0.017	213.2
p(day+wind) Ψ(.)	4	221.25	6.87	0.016	213.3
p(time2) Ψ(.)	4	221.32	6.94	0.016	213.3
p(day2) Ψ(.)	4	221.36	6.98	0.015	213.4
p(temp*wind) Ψ(.)	5	222.43	8.05	0.009	212.4
p(time+precip+wind) Ψ(.)	5	222.90	8.52	0.007	212.9

[a] Number of parameters; [b] Akaike's Information Criterion; [c] Difference in the model's AIC from that of the top model; [d] Akaike weight. (.) Indicates that the parameter Ψ was held constant.

We detected Rusty Blackbirds at 35 out of 60 sites. Based on the null occupancy model without site covariates, Rusty Blackbirds used 0.629 ± 0.07 SE of the study sites. With $\alpha = 0.05$ and a c-hat value of less than 3, we concluded that our global wetland model had an acceptable fit [44]. However, because a c-hat value greater than 1 suggests overdispersion, we adjusted standard error estimates for each wetland use model by a factor of c-hat [44] and ranked models based on Quasi Akaike's Information Criterion (QAIC) scores.

The top model (number of parameters k = 7, −2 log-likelihood = 188.231, QAIC = 123.4366), included the survey covariate "visit" in the detection probability linear predictor and the site covariates "invertebrate abundance" and "mud" in the occupancy linear predictor (Table 3). Rusty Blackbirds preferred sites with higher aquatic invertebrate abundance (Figure 2a) and lower percent mud cover (Figure 2b). This model accounted for over 60% of the adjusted model weight. Because the second model was not within four delta QAIC units of the top model, we did not model average parameter estimates across all of the models included in the candidate set of wetland use models [45].

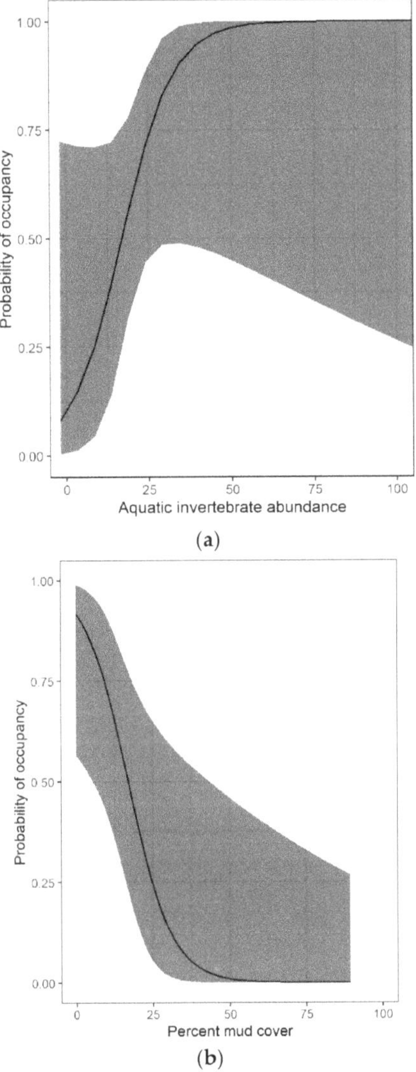

Figure 2. Relationship between probability of Rusty Blackbird occupancy and site covariates, percent cover of aquatic invertebrate abundance (**a**) and mud (**b**), with 95% confidence bands.

Aquatic invertebrate richness (across all taxa) was not a predictor of Rusty Blackbird wetland use. However, it is worth noting the extent to which each taxon varied in abundance across our study sites as well as the difference (or lack thereof) between taxa at sites with and without Rusty Blackbird detections. Survey sites had a mean of 6.92 insect families ($\pm$0.37 SE; range = 1 to 14) and a mean invertebrate count of 49.98 specimens ($\pm$6.70 SE; range = 6 to 205.5) combined from three samples in 2014. Sites with Rusty Blackbird detections had higher maximum invertebrate abundance of Amphipoda ($p < 0.001$), Coleoptera ($p = 0.002$), Collembola ($p < 0.001$), Diptera ($p < 0.001$), Hemiptera ($p = 0.018$), Odonata ($p < 0.001$), Oligochaeta ($p = 0.013$), Plecoptera ($p = 0.004$), and Trichoptera ($p = 0.033$) than did sites with no detections (Table 4).

Table 3. Model selection for wetland use of Rusty Blackbirds in northern New England in 2014, using site occupancy analysis with AIC scores adjusted for overdispersion (c-hat = 1.78).

Model	K^a	QAIC [b]	ΔQAIC [c]	Qw_i [d]	-2 Log-Likelihood
p(visit) Ψ(invert.abundance+mud)	7	123.44	0.00	0.600	188.2
p(visit) Ψ(mud)	6	128.06	4.63	0.060	199.6
p(visit) Ψ(invert.abundance)	6	128.31	4.87	0.053	200.0
p(visit) Ψ(invert.abundance+yng.conifer)	7	129.87	6.43	0.024	199.3
p(visit) Ψ(.)	5	129.99	6.55	0.023	206.4
p(visit) Ψ(mud+yng.conifer)	7	130.05	6.61	0.022	199.6
p(visit) Ψ(beaver+invert.abundance)	7	130.12	6.69	0.021	199.7
p(.) Ψ(.)	3	130.21	6.77	0.020	213.6
p(visit) Ψ(invert.abundance+pct.conifer)	7	130.35	6.92	0.019	200.1
p(visit) Ψ(invert.richness)	6	130.73	7.29	0.016	204.2
p(visit) Ψ(open.water)	6	130.94	7.51	0.014	204.6
p(visit) Ψ(pct.conifer)	6	131.07	7.63	0.013	204.8
p(visit) Ψ(yng.sof+beaver+invert.abundance)	8	131.58	8.14	0.010	198.8
p(visit) Ψ(yng.conifer)	6	131.91	8.47	0.009	206.2
p(visit) Ψ(puddles3x)	6	131.92	8.49	0.009	206.3
p(visit) Ψ(elevation)	6	131.93	8.49	0.009	206.3
p(visit) Ψ(size)	6	131.97	8.53	0.008	206.3
p(visit) Ψ(water.depth)	6	131.97	8.53	0.008	206.4
p(visit) Ψ(beaver)	6	131.99	8.55	0.008	206.4
p(visit) Ψ(invert.richness+yng.conifer)	7	132.68	9.24	0.006	204.1
p(visit) Ψ(beaver+invert.richness)	7	132.73	9.29	0.006	204.2
p(visit) Ψ(open.water+pct.conifer)	7	132.91	9.48	0.005	204.5
p(visit) Ψ(pct.conifer+puddles.3x)	7	132.97	9.53	0.005	204.6
p(visit) Ψ(pct.conifer+beaver)	7	133.05	9.61	0.005	204.8
p(visit) Ψ(yng.conifer)	7	133.84	10.41	0.003	206.1
p(visit) Ψ(size+yng.conifer)	7	133.86	10.43	0.003	206.1
p(visit) Ψ(water.depth+yng.conifer)	7	133.89	10.46	0.003	206.2
p(visit) Ψ(yng.conifer+beaver)	7	133.90	10.46	0.003	206.2
p(visit) Ψ(water.depth+size)	7	133.95	10.52	0.003	206.3
p(visit) Ψ(beaver+size)	7	133.96	10.53	0.003	206.3
p(visit) Ψ(pct.conifer+beaver+puddles.3x)	8	134.97	11.53	0.002	204.6

[a] Number of parameters. [b] Quasi Akaike's Information Criterion. [c] Difference in the model's QAIC from that of the top model. [d] Quasi Akaike weight.

Table 4. Summary statistics and results of a 2-sample Poisson rate test for differences between maximum invertebrate specimen abundance per survey per order from three aquatic macroinvertebrate surveys per site for sites with and without Rusty Blackbird (RUBL) detections in northern New England in 2014.

Order	RUBL Used [a]			RUBL Undetected [b]			Poisson Estimates			
	Max Count	N	Mean	Max Count	N	Mean	Estimate for Difference [c]	95% Lower Bound	Z	p-Value [d]
Amphipoda	112	35	3.20	28	25	1.12	2.08	1.47	5.64	<0.001
Aranae [e]	4	35	0.11	3	25	0.12	−0.01	−0.15	−0.06	0.525
Coleoptera	49	35	1.40	16	25	0.64	0.76	0.34	2.97	0.002
Collembola [e]	143	35	4.09	0	25	0.00	4.09	3.52	11.96	<0.001
Diptera	789	35	22.54	284	25	11.36	11.18	9.46	10.67	<0.001
Ephemeroptera	74	35	2.11	165	25	6.60	−4.49	−5.42	−7.88	1.000
Hemiptera [e]	8	35	0.23	1	25	0.04	0.19	0.04	2.09	0.018
Hirudinea [e, f]	1	35	0.03	6	25	0.24	−0.21	−0.38	−2.07	0.981
Lepidoptera [e]	2	35	0.06	0	25	0.00	0.06	−0.01	1.41	0.079
Megaloptera [e]	5	35	0.14	14	25	0.56	−0.41	−0.68	−2.56	0.995
Odonata	110	35	3.14	38	25	1.52	1.62	0.98	4.18	<0.001
Oligochaeta [e, f]	5	35	0.14	0	25	0.00	0.14	0.04	2.24	0.013
Plecoptera [e]	7	35	0.20	0	25	0.00	0.20	0.08	2.65	0.004
Trichoptera	74	35	2.11	37	25	1.48	0.63	0.07	1.83	0.033
Veneroida [e]	8	35	0.23	29	25	1.16	−0.93	−1.31	−4.05	1.000

[a] Considered used if at least one Rusty Blackbird was detected at least once during three surveys. [b] Considered undetected if no Rusty Blackbirds were detected during any of the three surveys. [c] Rate for RUBL-detected – RUBL-undetected sites. [d] Significant p-values (α = 0.05) are bolded. [e] The normal approximation may be inaccurate for small total number of occurrences. [f] Subclass, rather than order.

4. Discussion

4.1. Wetland Use and Aquatic Macroinvertebrate Prey

Food availability appears to be important to breeding Rusty Blackbird wetland use in northern New England. According to our model ranking, the best predictors of Rusty Blackbird use of wetlands in northern New Hampshire and western Maine are aquatic invertebrate abundance (positively related) and percent cover of mud (negatively related). Contrary to our hypotheses, current beaver activity, presence of puddles, and conifer cover were not important factors. Our data suggest that adult Rusty Blackbirds choose foraging sites based on aquatic invertebrate prey abundance. Abundance of amphipods, beetles, true flies, dragonflies, and caddis flies were higher in sites with Rusty Blackbird detections. This is expected because young birds cannot easily digest dry seeds [30], and invertebrates contain protein needed for chicks' growing skeletons. Aquatic insects are high-quality prey items, in part because many emergents (e.g., Odonata) are soft-bodied in the teneral or subimago stage, vulnerable to capture, and easily digested.

Our model results suggest that invertebrate abundance is more important than invertebrate richness in predicting Rusty Blackbird wetland use. Given the spatial and temporal variability of beaver-influenced wetland food resources within the landscape, it makes sense that Rusty Blackbirds would choose sites with large numbers of prey items rather than sites with a diverse selection of aquatic invertebrate taxa. From 2010 to 2012, the mean distance of Rusty Blackbird nests in this area to nearest wetland was 409.65 m (+/− 46.15 SE) and the maximum distance was 1347 m [46]. Because the foraging habitat in the study area is patchy and nests can be 400 m or more from foraging sites [37], we would expect Rusty Blackbirds to seek out sites with abundant food resources to minimize time spent foraging and provisioning. Home ranges of 13 telemetered adult Rusty Blackbirds in Maine averaged 37.5 ha and ranged from 4–179 ha [9], suggesting that foraging strategies in this heterogeneous environment must be flexible but efficient for successfully raising chicks.

We observed Rusty Blackbirds catching and provisioning multiple prey items at once. This behavior has also been documented in a camera trap study [47] in our survey area as well as in other field observations in Maine and New Hampshire [48] and nearby Vermont [25]. Optimal Foraging Theory predicts how an animal may decide where to forage, what to forage for, and for how long to forage based on maximizing energy gained from food items while minimizing searching and handling time [49]. In addition to nesting hundreds of meters from foraging sites, Rusty Blackbirds often forage in multiple wetlands throughout their home range [9]. They may provision more food items at a time with increasing distance from the nest to the foraging site to maximize energy gained versus that expended, as has been observed in related Icterids [30]. Regional differences in breeding Rusty Blackbird foraging strategies may exist; an Alaskan study found that adult Rusty Blackbirds usually fed chicks one large (>2 cm long) prey item at a time [50]. Thus, it is possible that Rusty Blackbirds in Alaska may operate under a different foraging strategy based on finding high-quality prey rather than minimizing energy spent foraging.

Although invertebrate richness was not a strong predictor of Rusty Blackbird site use in our study, Rusty Blackbirds were more likely to use sites with higher aquatic invertebrate abundance. We modeled a total count of all aquatic invertebrates per site rather than abundance broken down by order due to small sample sizes. Because we suspected that some orders are more important than others, we conducted an exploratory analysis of Rusty Blackbird site use and invertebrate abundance by order. The total number of individual Coleopterans, Collembolans, Dipterans, and Odonates was three times as high at wetlands used by Rusty Blackbirds, four times as high for Amphipods, and twice as high for Trichopterans (Table 4). However, as noted, sites used by Rusty Blackbirds also had higher Coleoptera richness, underscoring the likely importance of aquatic beetles in the bird's diet. Although summer diet information is scarce for Rusty Blackbirds, aquatic beetles can make up 10% or more of their diet in spring and may exceed 25% in some regions of the US [20].

Prey availability is not consistent over time due to differences in insect life histories, weather changes, and other factors. Rusty Blackbirds forage aerially for Odonates and other flying insects as well as hunt at the water's edge for aquatic prey. Orians [30] suggested that invertebrate prey availability is higher in warm, dry weather due to higher insect emergence rates. In 2014, the weather in our study area was generally favorable for Odonate emergence, as mean temperature was 20.8 °C ± 0.39 SE and it rained during just 11% of surveys. Odonate emergence rates are highest during mid to late morning [30]; however, we opportunistically observed breeding Rusty Blackbirds foraging throughout the day. This observation suggests they may be choosing other prey later in the day, although we did not quantify the relative foraging time budgets for aerial, emergent, and aquatic prey. Future studies should consider using a combination of aerial and aquatic sampling methods to target different invertebrate orders and life stages.

4.2. Wetland Characteristics

Mud was a variable in our top wetland use model, but puddles, conifer cover, and current beaver activity were not important predictors of Rusty Blackbird wetland use. In contrast, Powell et al. found that Rusty Blackbird occupancy in New England was best explained by the presence of puddles (i.e., shallow pools of standing water), conifer cover greater than 70%, and evidence of current beaver activity [41]. Although that study did not find strong support for the mud cover survey covariate, our survey methods differed; Powell et al. used a binary measure of mud presence or absence within a site [41], whereas we estimated the percent cover of mud at the wetland.

Our research indicates that Rusty Blackbirds forage in wetlands with abundant aquatic invertebrates and low percent cover of mud. Percent cover of mud is conversely related to that of open water, in which macroinvertebrates located between the water surface and the substrate are accessible to prey-seeking birds. Furthermore, the Rusty Blackbird bill ranges from 17.5 to 19 mm in length [51] and is not morphologically designed to probe for prey in deep mud. Rusty Blackbirds foraging in mud are likely procuring invertebrates on the surface. Studies [52,53] have also suggested that wintering Rusty Blackbirds prefer sites with shallow water. In addition, Wright et al. [54] found that during migration, the birds use forest edges with leaf litter and shallow water, likely to take advantage of the proximity of both arthropods and perches.

Microhabitat features are likely important; we observed Rusty Blackbirds foraging from the surface of deep (>1 m) water while standing on emergent debris. A Vermont study noted that Rusty Blackbirds forage in the water from debris or logs [25]. Other studies in New England suggested an unclear relationship between shallow water extent and Rusty Blackbird occupancy [55]. Breeding Rusty Blackbirds forage in fens and wet meadows [46], yet bird use and prey communities in these shallow-water ecosystems remain understudied. Future breeding-season wetland surveys should note the presence of emergent substrates in deep standing water, as such microhabitats give Rusty Blackbirds access to otherwise inaccessible foraging areas. Additionally, future researchers should assess the heterogeneity of invertebrate food availability within each wetland, including within mud as well as around edges of emergent substrates within deeper water.

Current beaver activity in wetlands did not strongly influence wetland use by Rusty Blackbirds in our study area, as the model with wetland use as a function of beaver occupancy ranked lower than the null model. All of our study sites had been modified by beavers, which may have affected model performance. While the relationship between current beaver activity and breeding Rusty Blackbird wetland use is still unclear, beavers create both breeding and foraging habitat by increasing conifer cover and by making ponds [13], many of which persist for years to decades [56]. Furthermore, beaver-impounded streams contain greater numbers of Odonates [18], and Anisoptera nymphs prefer dams of woody debris over other habitat types [57], so beavers may increase the abundance of preferred food for breeding Rusty Blackbirds. Previous research found that the presence of current beaver activity increased the probability of Rusty Blackbird occupancy [41], which is expected given that beavers are associated with improved habitat for aquatic inverte-brates [58,59]. Furthermore, aquatic invertebrate availability is likely related to water depth and vegetation cover along the wetland edge, which are factors that beavers influence indi-rectly, rather than to the actual presence of beavers [57]. The relationship between beaver occupancy and Rusty Blackbird wetland use is worthy of further study and refinement.

4.3. Landscape Factors

Glennon [60] suggested that climate change and habitat modification are the main contributors to declines of several boreal bird species including the Rusty Blackbird. Fur-thermore, climate change is affecting the hydrology and invertebrate communities of North American boreal wetlands [61]. Sánchez-Bayo et al. [62] noted that species in several aquatic taxa known to be prey for Rusty Blackbirds (e.g., Odonata) have declined or disap-peared from many sites in North America. While we did not study insect loss or forested wetland change, the implication for the Rusty Blackbird's breeding habitat and prey base at the species' southern range limit, and perhaps across its North American summer range, is sobering. Although current beaver occupancy was not significant in the model, the influence of beavers on wetland hydrology, heterogeneity, and aquatic invertebrate as-semblages is strong [58]. Land managers within the Rusty Blackbird's breeding range should, to the degree feasible, continue to manage the boreal forest landscape by allowing beaver populations to persist and include a mixture of forest stand ages in planning. It is important to allow beaver populations to continue to create impoundments in order to help invertebrate-rich wetlands persist in a changing climate [63].

4.4. Probability of Detection

We found that the most important predictor of detection probability for breeding Rusty Blackbirds was the visit (survey) period. As we hypothesized, the probability of detection given wetland use was highest during the second visit, when parents were rearing nestlings (28 May–10 June 2014). Rusty Blackbirds tend to be highly secretive and hard to detect during nest-building, egg-laying, and incubation. Once eggs hatch, adult Rusty Blackbirds become more vocal and more obvious as they frequently forage for food and rear their young. Soon after fledging, Rusty Blackbird broods tend to move away from their nesting areas and towards wetlands [37]. Thus, we designed our study to capture differences in breeding season behavior by surveying for Rusty Blackbirds in three survey periods that coincide with their breeding stages. To maximize breeding season detection, future studies could focus sampling effort on the chick-rearing period.

Time of day, date, wind, temperature, precipitation, and wetland size were not impor-tant predictors of Rusty Blackbird detectability. Additional factors, including vegetation cover within a wetland, noise created by running water, and anthropogenic noise, could have impacted detectability. During our study, a few sites were within earshot of logging operations, but most surveys were not noticeably impacted by anthropogenic sounds. There is also a need to compare detectability among multiple wetland types. No infor-mation exists on Rusty Blackbird occupancy of fens or wet meadows, yet the birds often forage in these shallow-water ecosystems [46]. Such information would better prepare land managers to survey areas that have not been previously studied. Lastly, although we defined a site as a wetland, our actual unit of measurement is the distance over which we were able to detect Rusty Blackbirds; however, we were unable to accurately quantify the distance at which we could hear Rusty Blackbird calls or songs at each site.

4.5. Considerations

Our single-season occupancy analysis provides a snapshot of Rusty Blackbird use of wetlands in New England; the study was designed to characterize differences between wetlands used by Rusty Blackbirds and wetlands that were unlikely to have hosted foraging birds. Due to limited time and resources, our study scope was defined as wetlands within 500 m of a road, which could have caused bias. Between-year variation in prey availability and habitat features were not examined in this study but are likely important, especially given changes in hydrology through time resulting from shifting beaver occupancy and precipitation patterns.

We sampled a small area (approximately 1 m^2) for aquatic invertebrates at the edge of each wetland because the entire wetland perimeter was not accessible due to flooding or areas of downed trees. With multiple invertebrate surveys in the same marked area of each site, we were able to compare temporal changes in invertebrate food availability within a site as well as compare food availability for Rusty Blackbirds across a range of wetlands. However, because we did not sample all Rusty Blackbird prey species (such as snails and spiders) or all life stages of prey species, our surveys provide a useful but incomplete picture of each site's invertebrate community structure.

5. Conclusions

Our research suggests that Rusty Blackbirds forage in wetlands with abundant aquatic invertebrates and low percent cover of mud, using sites with more open water and emergent vegetation. Conservation of Rusty Blackbird populations and the diverse invertebrate communities upon which these birds depend will require land managers and biologists to explore uncharted territory. Because habitat change, mercury pollution, and climate change are regional to global issues that are difficult to address, we recommend that decision-makers within the breeding range focus on maintaining and improving nesting and foraging habitat. It is important that land managers retain existing beaver populations and manage hydrology in the face of climate change [63]. We recommend continued bird population and wetland monitoring especially because the relationships between water level and prey availability are mediated by climate and will likely experience greater variance over time. If this region becomes more drought-prone, land managers could experiment with managing wetland hydrology to support adequate soil moisture during the growing season by increasing the size of existing wetlands and creating new ones, mimicking the work of beavers.

Because much of the Canadian breeding range has not been surveyed for Rusty Blackbirds, US and Canadian researchers can collaborate to fill information gaps and identify key areas in need of protection. We recommend long-term monitoring of wetland habitat and aquatic invertebrates in the Acadian forest. Land managers, both public and private, have an exciting opportunity to help maintain and improve breeding habitats for Rusty Blackbirds and other imperiled boreal species. Conservationists should expand on education and engagement initiatives, such as the Rusty Blackbird Migration Blitz, to increase the general public's awareness of and concern for this species.

Author Contributions: Conceptualization, S.M., A.P., C.F., J.C., and S.F.; methodology, A.P., S.M., J.C. and C.F.; software, A.P. and J.C.; validation, A.P., S.M., C.F., and J.C.; formal analysis, A.P. and J.C.; investigation, A.P.; resources, S.M. and C.F.; data curation, A.P.; writing—original draft preparation, A.P. and S.M.; writing—review and editing, A.P., S.M., C.F., J.C., and S.F.; visualization, A.P. and S.M.; supervision, S.M., A.P., and C.F.; project administration, S.M. and A.P.; funding acquisition, S.M. and A.P. All authors have read and agreed to the published version of the manuscript.

Funding: This research was funded by Umbagog National Wildlife Refuge and the SUNY College of Environmental Science and Forestry Graduate Student Association Research Grant. The initial pilot study in 2013 was supported by the Northeastern States Research Cooperative through funding made available by the USDA Forest Service. The conclusions and opinions in this paper are those of the authors and not the NSRC, the Forest Service, or the USDA.

Institutional Review Board Statement: Not applicable.

Informed Consent Statement: Not applicable.

Data Availability Statement: Data presented in this study are available on request from the corresponding authors.

Acknowledgments: Sean Flint, Ian Drew, and other US Fish and Wildlife Service personnel from Umbagog National Wildlife Refuge provided field housing, site access, and administrative support. Dan Hudnut of Wagner Forest Management, Ltd. provided site access. Devon Cote, Kelsey Schumacher, Amasa Fiske-White, and Thomas Ruland conducted surveys as field technicians. Donald Arthur processed and identified aquatic invertebrate samples. Patricia Wohner provided field research support. Luke Losada Powell provided statistical and technical support.

Conflicts of Interest: The authors declare no conflict of interest.

References

1. Niven, D.K.; Sauer, J.R.; Butcher, G.S.; Link, W.A. Christmas Bird Count provides insights into population change in land birds hat breed in the boreal forest. *Am. Birds* **2004**, *58*, 10–20.
2. Greenberg, R.; Demarest, D.W.; Matsuoka, S.M.; Mettke-Hofmann, C.; Evers, D.C.; Hamel, P.B.; Luscier, J.D.; Powell, L.L.; Shaw, D.; Avery, M.L.; et al. Understanding Declines in the Rusty Blackbird *(Euphagus carolinus)*. In *Studies in Avian Biology, Boreal Birds of North America: A Hemispheric View of Their Conservation Links and Significance*; Wells, J.V., Ed.; University of California Press: Berkeley, CA, USA, 2011; pp. 107–126.
3. Greenberg, R.; Matsuoka, S.M. Rangewide ecology of the declining Rusty Blackbird: Mysteries of a species in decline. *Condor* **2010**, *112*, 770–777. [CrossRef]
4. McClure, C.J.W.; Rolek, B.W.; McDonald, K.; Hill, G.E. Climate change and the decline of a once common bird: Climate change and blackbird decline. *Ecol. Evol.* **2012**, *2*, 370–378. [CrossRef]
5. U.S. Fish and Wildlife Service Migratory Bird Program Focal Species. Available online: https://www.fws.gov/birds/management/managed-species/focal-species.php (accessed on 9 May 2020).
6. BirdLife International. The IUCN Red List of Threatened Species, Euphagus carolinus. Available online: https://dx.doi.org/10.2305/IUCN.UK.20182.RLTS.T22724329A131889624.en (accessed on 9 May 2020).
7. Edmonds, S.T.; Evers, D.C.; Cristol, D.A.; Mettke-Hofmann, C.; Powell, L.L.; McGann, A.J.; Armiger, J.W.; Lane, O.P.; Tessler, D.F.; Newell, P.; et al. Geographic and seasonal variation in mercury exposure of the declining Rusty Blackbird. *Condor* **2010**, *112*, 789–799. [CrossRef]
8. Barnard, W.H.; Mettke-Hofmann, C.; Matsuoka, S.M. Prevalence of hematozoa infections among breeding and wintering Rusty Blackbirds. *Condor* **2010**, *112*, 849–853. [CrossRef]
9. Powell, L.L.; Hodgman, T.P.; Glanz, W.E. Home ranges of Rusty Blackbirds breeding in wetlands: How much would buffers from timber harvest protect habitat? *Condor* **2010**, *112*, 834–840. [CrossRef]
10. Hamel, P.B.; Steven, D.D.; Leininger, T.; Wilson, R. Historical trends in Rusty Blackbird nonbreeding habitat in forested wetlands. In Proceedings of the Fourth International Partners in Flight Conference: Tundra to Tropics, McAllen, TX, USA, 13–16 February 2009; Rich, T.D., Arizmendi, C., Demarest, D.W., Thompson, C., Eds.; pp. 341–353.
11. Luepold, S.H.B.; Hodgman, T.P.; McNulty, S.A.; Cohen, J.; Foss, C.R. Habitat selection, nest survival, and nest predators of Rusty Blackbirds in northern New England, USA. *Condor* **2015**, *117*, 609–623. [CrossRef]
12. Rosell, F.; Bozser, O.; Collen, P.; Parker, H. Ecological impact of beavers *Castor fiber* and *Castor canadensis* and their ability to modify ecosystems. *Mammal Rev.* **2005**, *35*, 248–276. [CrossRef]
13. Müller-Schwarze, D.; Sun, L. *The Beaver: Natural History of a Wetlands Engineer*; Cornell University Press: Ithaca, NY, USA, 2003.
14. Johnston, C.A.; Naiman, R.K. Browse selection by beaver: Effects on riparian forest composition. *Can. J. For. Res.* **1990**, *20*, 1036–1043. [CrossRef]
15. Hood, G.A.; Larson, D.G. Ecological engineering and aquatic connectivity: A new perspective from beaver-modified wetlands. *Freshw. Biol.* **2015**, *60*, 198–208. [CrossRef]
16. Cunningham, J.M.; Calhoun, A.J.K.; Glanz, W.E. Patterns of beaver colonization and wetland change in Acadia National Park. *Northeast. Nat.* **2006**, *13*, 583–596. [CrossRef]
17. Margolis, B.E.; Raesly, R.L.; Shumwaya, D.L. The effects of beaver-created wetlands on the benthic macroinvertebrate assemblages of two Appalachian streams. *Wetlands* **2001**, *21*, 554–563. [CrossRef]
18. McDowell, D.M.; Naiman, R.J. Structure and function of a benthic invertebrate stream community as influenced by beaver (*Castor canadensis*). *Oecologia* **1986**, *68*, 481–489. [CrossRef]
19. Beecher, W.J. Adaptations for food-getting in the American blackbirds. *Auk* **1951**, *68*, 411–440. [CrossRef]
20. Beal, F.E.L. *Food of the Bobolink, Blackbirds, and Grackles*; U.S. Department Agriculture Biological Survey Bulletin, 13: Washington, DC, USA, 1900.
21. Bent, A.C. *Life Histories of North American Blackbirds, Orioles, Tanagers, and Their Allies*; United States National Museum Bulletin: Washinton, DC, USA, 1958; p. 211.
22. Martin, A.C.; Zim, H.S.; Nelson, A.L. *American Wildlife & Plants: A Guide to Wildlife Food Habits*; Dover Publications, Inc.: New York, NY, USA, 1951; ISBN 978-0-486-20793-3.
23. Meanley, B. *Blackbirds and the Southern Rice Crop*; U.S. Department of the Interior, Fish and Wildlife Service: Washington, DC, USA, 1971.
24. Avery, M.L. Rusty Blackbird (*Euphagus carolinus*). In *The Birds of North America 2000*; Academy of Natural Sciences: Philadelphia, PA, USA; American Ornithologists' Union: Washington, DC, USA, 1995; p. 200. [CrossRef]

25. Ellison, W.G. *The status and habitat of the Rusty Blackbird in Caledonia and Essex Counties*; Vermont Fish and Wildlife Department: Woodstock, VT, USA, 1990.

26. Cade, T.J. Aerial feeding of the Rusty Blackbird on mosquitoes. *Wilson Bull.* **1953**, *65*, 52–53.

27. Ehrlich, P.; Dobkin, D.S.; Wheye, D. *Birder's Handbook*; Simon and Schuster: New York, NY, USA, 1988.

28. Matsuoka, S.M.; Shaw, D.; Johnson, J.A. Estimating the abundance of nesting Rusty Blackbirds in relation to wetland habitats in Alaska. *Condor* **2010**, *112*, 825–833. [CrossRef]

29. Edmonds, S.T.; O'Driscoll, N.J.; Hillier, N.K.; Atwood, J.L.; Evers, D.C. Factors regulating the bioavailability of methylmercury to breeding rusty blackbirds in northeastern wetlands. *Environ. Pollut.* **2012**, *171*, 148–154. [CrossRef] [PubMed]

30. Orians, G.H. *Blackbirds of the Americas*; University of Washington Press: Seattle, WA, USA, 1985.

31. MacKenzie, D.I.; Nichols, J.D.; Lachman, G.B.; Droege, S.; Royle, J.A.; Langtimm, C.A. Estimating site occupancy when detection probabilities are less than one. *Ecology* **2002**, *83*, 2248–2255. [CrossRef]

32. U.S. Climate Data. Available online: https://www.usclimatedata.com/climate/colebrook/new-hampshire/united-states/usnh0044 (accessed on 5 October 2020).

33. Pachomski, A.L. Foraging Habitat Characteristics, Prey Availability, and Detectability of Rusty Blackbirds: Implications for Land and Wildlife Management in the Northern Forest. Master's Thesis, State University of New York College of Environmental Science and Forestry, Syracuse, NY, USA, 2017.

34. Tavernia, B.G.; Lyons, J.E.; Loges, B.W.; Wilson, A.; Collazo, J.A.; Runge, M.C. An evaluation of rapid methods for monitoring vegetation characteristics of wetland bird habitat. *Wetl. Ecol. Manag.* **2015**, *23*, 495–505. [CrossRef]

35. Peckarsky, B.L.; Fraissinet, P.R.; Penton, M.; Conklin, D.D., Jr. *Freshwater Macroinvertebrates of Northeastern North America*; Cornell University Press: Ithaca, NY, USA, 1990.

36. Homer, C.G.; Dewitz, J.A.; Yang, L.; Jin, S.; Danielson, P.; Xian, G.; Coulston, J.; Herold, N.D.; Wickham, J.D.; Megown, K. Completion of the 2011 National Land Cover Database for the conterminous United States representing a decade of land cover change information. *Photogramm. Eng. Remote Sens.* **2015**, *81*, 345–354.

37. Wohner, P.J.; Foss, C.R.; Cooper, R.J. Rusty blackbird habitat selection and survivorship during nesting and post-fledging. *Diversity* **2020**, *12*, 221. [CrossRef]

38. Behnke, L.A.H.; Pejchar, L.; Crampton, L.H. Occupancy and habitat use of the endangered Akikiki and Akekee on Kauai Island, Hawaii. *Condor* **2016**, *118*, 148–158. [CrossRef]

39. Fiske, I.J.; Chandler, R.B. Unmarked: An R package for fitting hierarchical models of wildlife occurrence and abundance. *J. Stat. Softw.* **2011**, *43*, 1–23. Available online: http://www.jstatsoft.org/v43/i10/ (accessed on 14 August 2016). [CrossRef]

40. R Development Core Team. *R version 3.2.5: A Language and Environment for Statistical Computing*; R Foundation for Statistical Computing: Vienna, Austria, 2016; Available online: http://www.R-project.org (accessed on 14 August 2016).

41. Powell, L.L.; Hodgman, T.P.; Fiske, I.J.; Glanz, W.E. Habitat occupancy of Rusty Blackbirds (*Euphagus carolinus*) breeding in northern New England, USA. *Condor* **2014**, *116*, 122–133. [CrossRef]

42. Mazerolle, M.J. AICcmodavg: Model Selection and Multimodel Inference Based on (Q)AIC(c). R Package Version 2.0-4. 2016. Available online: http://CRAN.R-project.org/package=AICcmodavg (accessed on 14 August 2016).

43. MacKenzie, D.I.; Bailey, L.L. Assessing the fit of site-occupancy models. *J. Agric. Biol. Environ. Stat.* **2004**, *9*, 300–318. [CrossRef]

44. Lebreton, J.D.; Burnham, K.P.; Clobert, J.; Anderson, D.R. Modeling survival and testing biological hypotheses using marked animals: A unified approach with case studies. *Ecol. Monogr.* **1992**, *62*, 67–118. [CrossRef]

45. Burnham, K.P.; Anderson, D.R. *Model Selection and Multimodel Inference: A Practical Information-Theoretic Approach*, 2nd ed.; Springer-Verlag: New York, NY, USA, 2002.

46. Foss, C.; Wohner, P.J. Rusty Blackbird Nest Locations in Relation to Wetlands. Unpublished.

47. Buckley, S.H. Rusty Blackbirds in Northeastern U.S. Industrial Forests: A Multi-Scale Study of Nest Habitat Selection and Nest Survival. Master's Thesis, State University of New York College of Environmental Science and Forestry, Syracuse, NY, USA, 2013.

48. Foss, C. (New Hampshire Audubon, Concord, NH, USA). Personal communication, 2017.

49. Sinervo, B. Chapter 6: Optimal Foraging Theory: Constraints and Cognitive Processes. In *Behavioral Ecology*; University of California: Santa Cruz, CA, USA, 1997.

50. Loomis, D. Reproductive Success and Foraging Ecology of the Rusty Blackbird on the Copper River Delta, Alaska. Master's Thesis, Oregon State University, Corvallis, OR, USA, 2013.

51. Meanley, B. *Aging and Sexing Blackbirds, Bobolinks, and Starlings*; U.S. Fish Wildlife Service, Patuxent Wildlife Research Center: Laurel, MD, USA, 1967.

52. Borchert, S.M. Site-Specific Habitat and Landscape Associations of Rusty Blackbirds Wintering in Louisiana. Master's Thesis, Louisiana State University, Baton Rouge, LA, USA, 2015.

53. DeLeon, E.E. Ecology of Rusty Blackbirds Wintering in Louisiana: Seasonal Trends, Flock Composition, and Habitat Associations. Master's Thesis, Louisiana State University, Baton Rouge, LA, USA, 2012.

54. Wright, J.R.; Powell, L.L.; Matthews, S.N.; Tonra, C.M. Rusty blackbirds select areas of greater habitat complexity during stopover. *Condor* **2020**, *122*, 1–18. [CrossRef]

55. Scarl, J.C. *Rusty Blackbirds 2012: Building Connections for a Declining Species*; Vermont Center for Ecostudies: Norwich, VT, USA, 2013.

56. Naiman, R.J.; Johnston, C.A.; Kelley, J.C. Alteration of North American streams by beaver. *BioScience* **1988**, *38*, 753–762. [CrossRef]

57. Burcher, C.; Smock, L. Habitat distribution, dietary composition and life history characteristics of odonate nymphs in a blackwater coastal plain stream. *Am. Midl. Nat.* **2002**, *148*, 75–89. [CrossRef]
58. Hood, G.A.; Larson, D.G. Beaver-created habitat heterogeneity influences aquatic invertebrate assemblages in boreal Canada. *Wetlands* **2014**, *34*, 19–29. [CrossRef]
59. Bush, B.M.; Stenert, C.; Maltchik, L.; Batzer, D.P. Beaver-created successional gradients increase β-diversity of invertebrates by turnover in stream-wetland complexes. *Freshw. Biol.* **2019**, *64*, 1265–1274. [CrossRef]
60. Glennon, M. *Analysis of Boreal Bird Trends in the Adirondack Park, 2007–2016: Report to the New York State Department of Environmental Conservation*; Wildlife Conservation Society: Saranac Lake, NY, USA, 2017.
61. Corcoran, R.M.; Lovvorn, J.R.; Heglund, P.J. Long-term change in limnology and invertebrates in Alaskan boreal wetlands. *Hydrobiologia* **2009**, *620*, 77–89. [CrossRef]
62. Sánchez-Bayo, F.; Wyckhuys, K.A.G. Worldwide decline of the entomofauna: A review of its drivers. *Biol. Conserv.* **2019**, *232*, 8–27. [CrossRef]
63. Hood, G.A.; Bayley, S.E. Beaver (*Castor canadensis*) mitigate the effects of climate on the area of open water in boreal wetlands in western Canada. *Biol. Conserv.* **2008**, *141*, 556–557. [CrossRef]

Article

Rusty Blackbird Habitat Selection and Survivorship during Nesting and Post-Fledging

Patricia J. Wohner [1,*], **Carol R. Foss** [1] and **Robert J. Cooper** [2]

[1] New Hampshire Audubon, Concord, NH 03301, USA; cfoss@nhaudubon.org
[2] Warnell School of Forestry, University of Georgia, Athens, GA 30602, USA; rcooper@warnell.uga.edu
* Correspondence: pjwohner@gmail.com

Received: 13 May 2020; Accepted: 29 May 2020; Published: 2 June 2020

Abstract: Rusty blackbird (*Euphagus carolinus*) populations have declined dramatically since the 1970s and the cause of decline is still unclear. As is the case for many passerines, most research on rusty blackbirds occurs during the nesting period. Nest success is relatively high in most of the rusty blackbird's range, but survival during the post-fledging period, when fledgling songbirds are particularly vulnerable, has not been studied. We assessed fledgling and adult survivorship and nest success in northern New Hampshire from May to August in 2010 to 2012. We also assessed fledgling and adult post-fledging habitat selection and nest-site selection. The likelihood of rusty blackbirds nesting in a given area increased with an increasing proportion of softwood/mixed-wood sapling stands and decreasing distances to first to sixth order streams. Wetlands were not selected for nest sites, but both adults and fledglings selected wetlands for post-fledging habitat. Fledglings and adults selected similar habitat post-fledging, but fledglings were much more likely to be found in habitat with an increasing proportion of softwood/mixed-wood sapling stands and were more likely to be closer to streams than adults. No habitat variables selected during nesting or post-fledging influenced daily survival rates, which were relatively low for adults over the 60-day study periods (males 0.996, females 0.998). Fledgling survival rates (0.89) were much higher than reported for species of similar size.

Keywords: *Euphagus carolinus*; nest success; post-fledging; rusty blackbird; survivorship; streams; wetlands

---

1. Introduction

Songbird breeding productivity is often measured by nest success and associated habitat characteristics [1]. However, mortality during the post-fledging period can also result in low breeding productivity and limit populations even when nest success is high [2,3]. Juvenile songbird survivorship is typically estimated to be 50–60% of adult survivorship [4,5]. However, actual juvenile survival rate estimates are quite variable, ranging from 0.19 in hooded warblers (*Setophaga citrina*) to >0.65 in ovenbird (*Seiurus aurocapilla*) and worm-eating warbler (*Helmitheros vermivorum*; [1,3]. Despite this potential variability in survivorship, the post-fledging period remains an understudied component of avian demographics for many species in part due to difficulties of following individuals after fledging [5–9].

Differences in food requirements and vulnerability to predators between the nesting and post-fledging periods may lead to differences in habitat selection between these stages of the breeding season [7,10,11]. For example, ovenbird, worm-eating warbler, hooded warbler, Acadian flycatcher (*Empidonax virescens*), cerulean warbler, red-eyed vireos (*Vireo olivaceus*), and Swainson's thrush (*Catharus ustulatus*) families all move to stands with greater structural complexity following fledging [1,2,11–15]. High mortality rates are associated with this movement as young birds seek better-suited post-fledging habitat [6,7]. Therefore, full assessment of breeding habitat quality requires evaluation of post-fledging habitat selection and survival [16].

The rusty blackbird (*Euphagus carolinus*) is a continental migrant that breeds primarily in the boreal forests of Canada and Alaska, with populations in Acadian forests of northern New England and eastern Canada [17] and northern New York [18,19]. The species has experienced an estimated 85–95% continent-wide population decline with accelerated decline rates beginning in the early 1970s [20–24]. It has been declared a species of conservation concern by the United States Fish and Wildlife Service [25] and is considered vulnerable to extinction by the International Union for the Conservation of Nature [26]. Reasons for the decline remain unclear despite extensive study on both the breeding and wintering grounds. However, habitat use and survivorship during the post-fledging period have not been studied.

Nest success is relatively high where studied in Alaska ($\bar{x} = 0.56$ [27]) and New England ($\bar{x} = 0.62$ [28], $\bar{x} = 0.53$ [29]), and rusty blackbirds do not seem to be suffering from chronically low rates of nest success. Rusty blackbird nest success estimates are much higher than the 0.30–0.39 estimated for red-winged (*Agelaius phoeniceus*), yellow-headed (*Xanthocephalus xanthocephalus*), and Brewer's (*Euphagus cyanocephalus*) blackbirds, which have not experienced dramatic declines [30].

Rusty blackbirds typically nest in young spruce-fir stands with high stem densities and forage for aquatic invertebrates such as Ephemeroptera, Odonata, Plecoptera, Diptera, and Tricoptera [17,31] in shallow wetlands or along stream edges [17,28,29,31]. Previous studies have identified percent cover of young softwood stands to be the most important variable selected for nest sites at multiple scales and associated with higher survival [27–29]. However, wetlands were not selected for nesting at the 5 m or 500 m radius scale even though the species is well-known as a wetland obligate [17].

Understanding the roles of both wetlands and young softwood stands could have important consequences for research and management. Dense spruce-fir sapling stands are selected for concealment during nesting and the first few days after fledging. Similar to many songbird species, young rusty blackbird fledglings fly poorly and lack the ability to escape predation in the first week or so after leaving the nest [8]. Before they become capable of sustained flight, fledglings spend much of their time on or near the ground, where they are easy prey for predators [2,3,7,10,32]. As energetic demands increase [33] and fledglings become more mobile, proximity to wetlands and streams with abundant and diverse invertebrate resources becomes important as fledglings learn to forage independently and fulfill their high energetic demands [6,7,31,33].

Our objectives were to 1) determine what forest stand types are important to rusty blackbirds during nesting and post-fledging, 2) assess nest site and post-fledging habitat selection, 3) analyze the influence of selected habitat characteristics on nest success and post-fledging survivorship of fledglings and adults, 4) provide forest management recommendations for incorporating post-fledging habitat into management plans for rusty blackbirds, and 5) advocate a paradigm shift from wetland-centric to landscape-centric for future rusty blackbird researchers in New England, New York, and Maritime Canada.

2. Materials and Methods

2.1. Study Area

We studied rusty blackbirds during 2010–2012 in the upper Androscoggin River watershed in northern New Hampshire. Ecosystem classification places this watershed within the White Mountains Section of the New England-Adirondack Province [34]. The landscape is mountainous, with most of the area at elevations between 460 to 800 m.a.s.l., valleys between 305 to 460 m.a.s.l., and a few peaks and ridges exceeding 800 m.a.s.l. [35]. Surface waters include the 6th order Androscoggin and Magalloway rivers, 4th order Clear Stream and Swift Diamond River, numerous lower order streams, Umbagog Lake (3177 ha), and several ponds of 10 to 125 ha. Climatic conditions include cold winters and warm summers, with mean monthly lows ranging from −15 °C to 13 °C and highs from −3.3 °C to

25.6 °C in January and July, respectively; mean annual precipitation is 105 cm, with a monthly means ranging from 5.9 cm in February to 11.1 cm in October, and 198 cm of snowfall [36].

The area is heavily forested in a patchwork of northern hardwood, Acadian spruce-fir, and mixed northern hardwood-spruce-fir. Balsam fir (*Abies balsamea*) and red spruce (*Picea rubens*) dominate softwood stands; eastern white pine (*Pinus strobus*), black spruce (*Picea mariana*), northern white cedar (*Thuja occidentalis*), and tamarack (*Larix laricina*) occur at lower densities. Major hardwood species include sugar maple (*Acer saccarum*), red maple (*Acer rubrum*), yellow birch (*Betula alleghaniensis*), white birch (*Betula papyrifera*), quaking aspen (*Populus tremuloides*), bigtooth aspen (*Populus grandidentata*), balsam poplar (*Populus balsamifera*), American beech (*Fagus grandifolia*), black cherry (*Prunus serotina*), pin cherry (*Prunus pensylvanica*), American mountain-ash (*Sorbus americana*), striped maple (*Acer pensylvanicum*), white ash (*Fraxinus americana*), and black ash (*Fraxinus nigra*). Primary stand disturbance agents include timber harvesting, wind throw, breakage from ice and snow loads, beavers (*Castor canadensis*), and insect outbreaks, notably spruce budworm (*Choristoneura* spp.) [35].

We conducted our study in three drainages: 1) Swift Diamond River valley (SWDI), 2) Mollidgewock (MOLL), and 3) Interior (INTE). SWDI was located on Wagner Forest Management Ltd. lands (hereafter Wagner) and INTE and MOLL were located on the Umbagog National Wildlife Refuge (hereafter Umbagog; Figure 1).

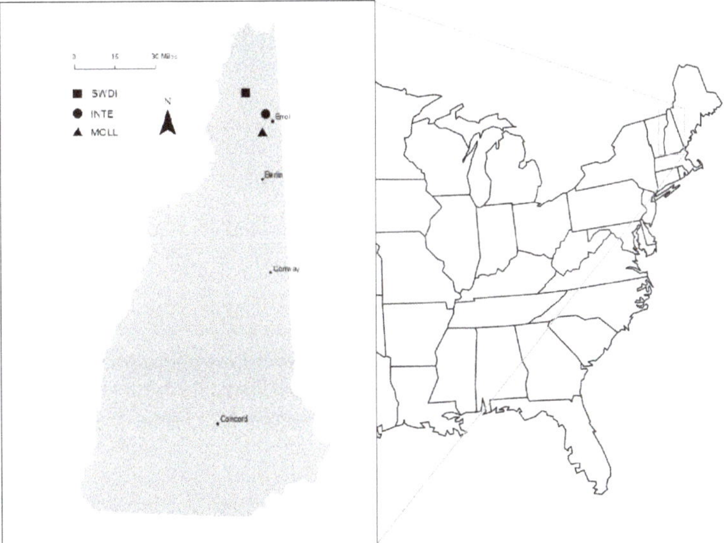

Figure 1. Drainages and their locations relative to the nearest township of Errol, Coos County, New Hampshire within the USA. Symbols indicate approximate site locations from 2010–2012 and include Swift Diamond (SWDI), Interior (INTE), and Mollidgewock (MOLL).

2.2. Field Procedures

During our pilot study in 2010, we conducted presence-absence surveys within 100 m of wetlands by observing passively for 3 min, broadcasting male rusty blackbird songs and calls for 38 s, and observing passively for another 5 min [28]. We discovered that these surveys failed to detect previously located rusty blackbird pairs that were nesting > 100 m from wetlands. Thus, we changed our survey protocol for 2011 and 2012 to include regenerating softwood stands up to approximately 0.5 km from wetlands during 30 min to 2 h passive surveys [27]. We identified survey locations using a combination of forest industry maps, topographic maps, and Google Earth imagery. We began nest-searching in early May in areas of rusty blackbird activity and continued through mid-June.

Once located, nests were monitored approximately every 5 days until they either fledged young or failed. When nests failed during incubation, we attempted to locate re-nests.

To collect spatial and survival data on fledgling and adult rusty blackbirds, we deployed radio-transmitters on a subset of nestling and adult rusty blackbirds captured at nest sites; not all nests received transmitters. We attached VHF transmitters to nestlings when they were approximately 7–10 days old. We captured adult rusty blackbirds near active nests using 60-mm mesh size, 6 m long mist nests, and a broadcast call. We used transmitters from Blackburn Transmitters (Nacogdoches, TX, USA), Advanced Telemetry Systems (Isanti, MN, USA), and Holohil Systems, Ltd. (Carp, ONT, CA). Battery life ranged from 30 days in 2010 to 60 days in 2012. Transmitters were attached via a synsacrum harness with a degradable 1 mm stretchy jelly cord [37]. In 2012, we designed a harness with a weak link because birds returned with harnesses from previous years that had become embedded in their skin.

Transmitter weight with harness varied from 1.8% (70 g bird) to 2.3% (55 g bird) of adult bird mass. Transmitters on nestlings weighed from 5.6 to 8% of mass at time of attachment but were lower than adult ratios once nestlings fledged because transmitters used on fledglings were lighter (0.9 g) than those for adults (2.7 g). Harnesses were fitted loosely on nestlings and were the largest size to enable growth. Each adult and nestling blackbird received a US Geological Survey (USGS) federal band and adults received a unique color-band combination for identification (USGS BBL Permit # 22665). This study plan was approved by the University of Georgia IACUC (AUP# A2009 1-003).

Observers located tagged rusty blackbirds 3–5 times per week using ATS R2000 and R2100 receivers in vehicles with roof-mounted dipole antennas and hand-held R410 receivers with three-element Yagi antennas. In > 94% of locations, we determined coordinates with a Garmin GPS by following a radio signal until we saw or heard the bird. In remaining cases, we used triangulation of bearings taken from multiple points on logging roads. We conducted observations between sunrise and sunset and avoided periods of excessive wind and rain.

2.3. Data Analyses

2.3.1. Habitat Selection

We used binomial logistic regression to determine habitat characteristics that rusty blackbirds used out of proportion to availability [38]. We conducted separate analyses for nests, fledgling, and adult telemetry locations. Nests or telemetry locations were considered use points and we generated random availability points within rusty blackbird areas of use with ARCMAP 10.4. We then assigned a value of 1 to use points and 0 to availability points to be used as response variables. If multiple tagged birds were documented at the same location at the same time, we only used a location once for a use point and prioritized triangulated points for removal.

To generate availability points, we created a 90% kernel density background for each of the three analyses and the three drainages separately. For the post-fledging analysis, the background area of use included area outside nesting sites because fledglings moved increasing distances (>1 km) from the nest as they became more mobile. We generated roughly the same number of availability points as use points. We constrained availability points to be at least 50 m apart for telemetry data and 300 m apart for nests. Availability points within large bodies of open water or other unsuitable habitats (e.g., paved road surfaces, buildings) were deleted and not replaced.

We created predictor habitat variables based on published literature [24–26,38], our field observations, habitat characteristics used in our revised survey protocol, and relevance to forest management. We generated predictor variables for SWDI from layers we created in ARCMAP 10.4 using forest stand maps (Wagner), and soil, river, and wetland maps from New Hampshire Geographically Referenced Analysis and Information Transfer System (NH GRANIT, hereafter GRANIT; Table 1). For MOLL and INTE, where forest stand maps were unavailable, we used aerial photography to digitize forest stand types within a 30 m radius around use and paired availability points for each habitat variable for every other telemetry point/bird/day.

Table 1. Descriptions of habitat variables used for generating logistic regression models for rusty blackbird nests and post-fledging adult and fledgling locations in northern New Hampshire from 2010–2012. Buffer is a 30 m radius circle around location points.

Variable	Description
lowdrainage	% of buffer in class poorly and very poorly drained
lowslope	% of buffer that is 1–8% slope
stream123	mean distance of buffer to 1st–3rd order streams
stream3456	mean distance of buffer to 3rd–6th order streams
allstreams	mean distance of buffer to 1st–6th order streams
softwood	% softwood stands in buffer
swsaplings [1]	% softwood or mixed softwood (>75% SW)
mixage	% mixed-age stands in buffer
wetlands	% any vegetated palustrine wetland in buffer
alder	% alder wetland in buffer
beaver	% beaver pond wetland buffer
forestwl	% forested wetland in buffer
seasonfl	% seasonally flooded wetland in buffer

[1] swsaplings were stands <10 yrs old or with saplings 1.3–14 cm DBH.

Once we created predictor variables, we extracted raster data for every other point/day/bird (due to time and computing constraints and to match Umbagog data) for the 13 variables (Table 1). We calculated values for predictor variables as the mean of raster cell values in a 30 m radius buffer around use and availability points. The 30 m scale is roughly an order of magnitude between the 5 and 500 m scales used as buffers for variables in previous research [28,29]. To summarize variables, we calculated the mean and standard deviation of use and availability points for nests and blackbird locations.

We used generalized linear mixed models (GLMM; R package lme4) in program R 3.4.3 [39] to run logistic nest site and post-fledging habitat selection analyses [40]. We determined Pearson correlations on all pairs of predictor variables prior to modeling and avoided including variables correlated at r > 0.40 in the same model. We used Akaike information criterion (AICc) to compare model fit between nested models with and without year as a random effect and determined that year did not account for significant variation. For post-fledging telemetry data, we included individual birds nested within drainage (SWDI, INTE, MOLL) as a random effect in each model to control for spatial autocorrelation [40]. Availability points were assigned to individual birds based on location. We did not consider pseudoreplication to be a problem because blackbirds did not maintain a consistent home range post-fledging. We did not include fledglings and adults from the same nest in the same analysis, randomly omitting one.

We used model selection to evaluate the relative plausibility of habitat selection models with combinations of the 13 predictor variables [41,42]. To avoid over-fitting models with parameters, we first reduced our set of 13 variables by comparing correlated sets individually with the second-order AICc. We retained variables for further analysis that had the lowest AICc scores and were better than the null model. We included one stream, wetland, and forest stand type variable in a model at a time to avoid correlation and then compared 17 candidate models. The top habitat selection models for males and females included the same variables with similar weights, and parameter estimates were similar, so we pooled males and females into adults for habitat selection analyses.

We assessed the relative fit of each candidate model by calculating Akaike weights (ωi; [42]. We report a confidence set of models that included only those candidate models with Akaike weights that were within 10% of the largest weight [43] for evaluating strength of evidence. If more than 3 models were within 10% of the model with the greatest ωi, we report the subset of models with $\omega i >$ 0.01. To choose the best model in analyses that generated several top models, we chose the model with the fewest parameters within ΔAICc < 2 of the top model. We chose the top model if the level of support ($\omega i > 0.6$) indicated that it was far better than other models.

We used the chi-square G statistic to evaluate the strength of our top models in explaining goodness-of-fit ($p < 0.05$; [44]). We assessed the predictive ability of our top models by estimating the Area Under the Curve (AUC) of the Receiver Operating Characteristic (ROC) curve, i.e., the rate of true positives (sensitivity) to false positives (1-specificity). We plotted ROC curves with the R package "pROC". We used an AUC ≥ 0.85 to indicate good predictive ability of our models.

We assessed the precision of model-averaged coefficients by calculating 95% confidence intervals. Confidence intervals including 1 indicated inconclusive estimates because we could not determine the nature of the relationships (i.e., whether positive or negative) due to imprecision in parameter estimates [45]. We report scaled, model-averaged coefficients from the top models ($\omega i > 0.6$). The scalars correspond to biologically relevant unit changes in predictors.

2.3.2. Nest Success and Adult and Fledgling Survivorship

We used the R package RMark 2.2.4 [46] to estimate daily survival for nests, fledglings, and adults. We used nest success models appropriate for use with telemetry data when data are not collected at a consistent time interval and the exact date of fates are not known [47,48]. For nest success, we considered the first day we observed eggs, incubation behavior, or nestlings to be the date the nest was found. We assumed a 29-day exposure time when calculating nest success; 5 days for laying, 12 days for incubation, and 12 days for nestlings [28]. We considered the date of fledging to be the "date found" in RMark for both adults and fledglings. Tagged nestlings that were depredated before they fledged were not included.

Covariates tested included variation in survival among years (2010, 2011, and 2012), among drainages (SWDI, MOLL, INTE), and with number of days since nest initiation (nests). To determine fledgling survival in the first week, we assessed seasonal survival of fledglings (season). We did not include adult age as a covariate because only two adults could be aged beyond the after-hatch-year class. We included habitat variables from top models from nest and post-fledging habitat selection analyses. We used overall means to interpolate habitat variables for 9 blackbirds that were missing habitat data to enable daily survival rate calculations for those individuals. To account for classes with no mortality, i.e., survivorship = 1 and variance cannot be estimated, we added a fictitious bird to the dataset in that class and assumed the bird died on the last occasion [28]. We assigned mean habitat covariate values to this bird. We calculated the mean of all use locations for each habitat variable to include in the corresponding survival record for each bird and assumed the mean was representative of variables leading up to predation or survival. We used a model selection approach to determine the best model for nest success and blackbird survival, such as the methods described above for habitat selection.

3. Results

3.1. General

We found 59 rusty blackbird nests during the 2010–2012 breeding seasons, the majority of which (92%) were in previously clear-cut, regenerating softwood-dominated stands 5–15 years post-harvest. These stands averaged 5.4 ha and ranged from 0.36–13 ha. Of the 59 nests in our study, 31 (53%) were within 75 m and 10 (17%) were > 200 m from a stream or wetland.

We used 37 nest locations that were > 100 m apart and generated 57 random availability points to use in logistic regression. We attached 36 radio-transmitters to adults (13 female, 23 male) and 25 to nestlings. After removing adult birds tagged from the same nests, we were left with points from 11 females and 18 males. Of the 25 nestlings tagged, 8 died before fledging. In the seven cases where two nestlings from a nest had transmitters, one of the two was omitted from the fledgling analysis, leaving 10 fledglings in the habitat selection analysis. We documented a total of 318 use locations for 20 blackbirds in 2010, 993 use locations for 24 birds in 2011, and 626 use locations for 17 birds in 2012 (Appendix A). We omitted location points for one female who died from apparently capture-related causes. We omitted 70 location points of adults and fledglings documented when fledglings were still in the nest to eliminate data from the nestling period. After refining the telemetry data, 771 use points remained, and we generated 806 availability points.

New fledglings flew weakly and most remained within 100 m of the nest for the first few days after fledging. Once fledglings moved away from the nesting area, we observed pairs dividing their fledglings between them, joining other family groups, and feeding fledglings of other pairs. Most fledglings depended on adults for food for about 3 weeks. Some fledglings traveled alone after becoming independent, while others remained in family groups or joined up with fledglings from their own or other families. During our study, all tagged individuals that fledged before 8 June had been detected more than 1 km from the nest by the end of the month.

3.2. Habitat Selection

The top model for habitat selection included similar variables for all 3 groups tested (i.e., nest sites, fledglings, and adults): distance to streams, proportion of softwood/mixed-wood sapling stands, wetlands, and low slope class (Table 2). In each case, the top model received strong support ($wi > 0.62$; Table 2). However, parameter estimates differed substantially among the groups, justifying not pooling adults and fledglings in the same analysis (Table 3).

Table 2. High ranking candidate models for rusty blackbird **nesting**, **fledgling**, and **adult** habitat selection in Northern New Hampshire from 2010–2012. k = number parameters, AICc = Akaike information criterion for small sample sizes, ΔAICc is the difference between the model and top model, wi = weight, cum., wi = cumulative weight, and LL = log likelihood. See Table 1 for variable definitions.

Candidate Models	k	AICc	ΔAICc	wi	cum. wi	LL
Nesting						
allstreams+swsaplings+lowslope	5	93.28	0	0.62	0.62	−41.3
allstreams+swsaplings+lowslope+wetlands	6	95.24	1.95	0.23	0.85	−41.14
stream123+swsaplings+lowslope+wetlands	6	96.81	3.53	0.11	0.96	−41.92
allstreams+swsaplings+wetlands	5	99.54	6.25	0.03	0.99	−44.43
Null	2	130.16	36.87	0	1	−63.01
Fledgling						
allstreams+swsaplings+lowslope+wetlands	6	404.91	0	0.73	0.73	−196.35
allstreams+swsaplings+lowslope+wetlands+softwood	7	406.97	2.06	0.26	0.99	−196.34
Null	3	543.62	138.71	0	1	−268.78
Adult						
allstreams+swsaplings+lowslope+wetlands	6	1164.69	0	0.73	0.73	−576.31
allstreams+swsaplings+lowslope+wetlands+softwood	7	1166.68	1.99	0.27	0.99	−576.29
allstreams+mixage+lowslope+wetlands+softwood	7	1174.46	9.77	0.01	1	−580.18
Null	3	1568.82	404.13	0	1	−781.40

Distance to 1st–6th order streams was an important variable in all 3 analyses, and had the largest effect for fledglings (Figure 2, Table 3). Fledglings were 4.6 times more likely to occur with every 50 m decrease in distance from streams, compared to 1.2 times more likely for adults and 1.3 times for nest sites. Nests, fledglings, and adults were all 1.2 times more likely to occur with every 10% increase in softwood/mixed-wood sapling stands. Only nests were influenced by low slope and were 1.1 times more likely to be observed with a 10% increase in area with low slope. While the proportion of wetlands in an area was important for fledgling and adult habitat use, (blackbirds were 1.3 times more likely with each 10% increase), wetlands were not important for nest-site selection (Figure 2).

Table 3. Adjusted, scaled back-transformed log-odds estimates for rusty blackbird **fledglings** (n = 10), **adults** (n = 29), and **nests** (n = 37) in New Hampshire from 2010–2012 including estimates, SE, and 95% upper and lower confidence intervals. Scaled log odds estimate reads: it is 1.3 times more likely a rusty blackbird will be present with every 10% increase in the proportion of sapling stands in a 30 m radius buffer. Confidence limits including 1 indicate no effect and are denoted "–".

Variable	Estimate	SE	LCI	UCI	Unit Change	Scaled Estimate	Scaled CL
Fledglings							
Intercept	−0.92	0.38	−1.66	−0.18			
distance to streams (1st–6th)	−30.3	7.76	−45.5	−15	50m	4.55	2.1, 9.7
proportion swsaplings	1.8	0.34	1.1	2.4	10%	1.2	1.1, 1.3
proportion low slope (1–8%)	1.18	0.3	0.59	1.77	–	–	
proportion wetland	2.9	0.6	1.71	4	10%	1.33	1.2, 1.5
Adults							
Intercept	−0.33	0.15	−0.62	−0.04			
distance to streams (1st–6th)	−3.7	0.4	−4.4	−2.9	50m	1.2	1.2, 1.3
proportion swsaplings	1.2	0.2	0.83	1.56	–	–	
proportion low slope (1–8%)	0.87	0.2	0.56	1.17	–	–	
proportion wetland	2.7	0.4	1.95	3.38	10%	1.31	1.2, 1.4
Nests							
Intercept	−0.87	0.55	−1.95	0.21			
distance to streams (1st–6th)	−5.16	1.65	−8.39	−1.93	50m	1.3	1.1, 1.5
proportion swsaplings	0.02	0.01	0.01	0.04	10%	1.2	1.1, 1.5
proportion low slope (1–8%)	0.01	0.01	0	0.03	10%	1.1	1, 1.3
proportion wetland	0.01	0.01	−0.02	0.04	–	–	

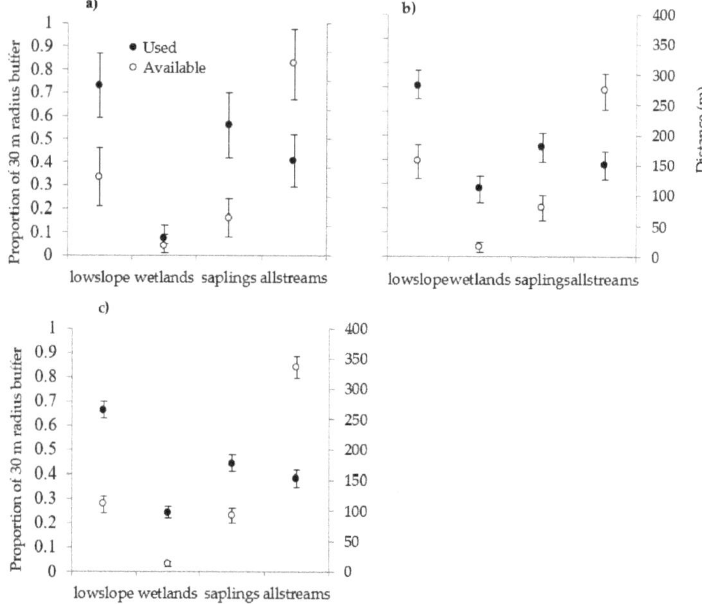

Figure 2. Mean and 95% CI of variables from top models of (**a**) nest use (black circles; n = 37) relative to availability (white circles; n = 57), (**b**) fledgling habitat use (black circles; n = 198) relative to availability (white circles; n = 196), and (**c**) adult use (black circles; n = 537) relative to availability (white circles; n = 609) in N. New Hampshire from 2010–2012. Proportion of lowslope, wetlands, and softwood/mixed-wood saplings are on the left *y*-axis, and distance (m) of points from allstreams is on the right *x*-axis. Lowslope = % of 30 m buffer in soil class poorly or very poorly drained, wetlands = % of 30 m buffer in any vegetated palustrine wetland, saplings = % of 30 m buffer in softwood/mixed-wood sapling stands, and allstreams = mean distance of buffer to 1st–6th order streams.

3.3. Nest Success

We included 37 nests with known outcomes that were > 100 m apart in the nest success analysis. The null model for nest survival received most of the support (Table 4); candidate models including year, drainage, and trend through time (season) received little support (Table 4). Daily nest survival for all years combined was 0.975, SE = 0.26, resulting in an overall nest success rate with 29-day exposure of 0.48 (Table 4). Although the top model for nest success included wetlands (Table 4), the confidence limits for the estimate of this variable included 0. We therefore chose the intercept only model as the top model and report estimates from this model (Table 5).

Table 4. High-ranking candidate models for rusty blackbird **fledgling, adult**, and **nest** survival in northern New Hampshire. k = number of parameters, AICc = Akaike information criterion adjusted for small sample sizes, ΔAICc is the difference between the model and the top model, ωi = weight, and cum. ωi = cumulative weight; see Table 1 for habitat variable descriptions. nestage is the age of the nest when first found, S = survival, time = time from first day of study, hy = fledgling, SWDI = Swift Diamond drainage.

Candidate Models	k	AICc	ΔAICc	ωi	Deviance
Nests					
S(~wetlands)	2	72.8	0	0.19	68.76
S(~nestage + wetlands)	3	73.5	0.69	0.13	67.42
S(~nestage + wetlands + allstreams)	4	74.3	1.55	0.09	66.24
S(~allstreams)	2	74.5	1.72	0.08	70.48
S(~wetlands + lowslope)	3	74.8	2.00	0.07	68.73
S(~nestage + wetlands + lowslope)	4	75.1	2.26	0.06	66.95
S(~1)	1	75.2	2.38	0.06	73.16
S(~nestage + allstreams)	3	75.9	3.07	0.04	69.8
S(~nestage)	2	76.0	3.16	0.04	71.92
S(~allstreams + swsaplings)	3	76.3	3.55	0.03	70.28
Adult males and females and fledglings (hy)					
S(~swsaplings + hy + female + male)	5	100.2	0	0.32	90.19
S(~hy + female + male)	4	102.0	1.8	0.13	94.02
S(~lowslope + allstreams)	3	102.9	2.6	0.09	96.84
S(~time + hy + female + male)	5	103.3	3.1	0.07	93.29
S(~lowslope + hy + female + male)	5	103.9	3.7	0.057	93.88
S(~lowslope)	2	104.0	3.8	0.05	99.98
S(~1)	1	104.1	3.9	0.05	102.09
S(~SWDI)	2	104.1	3.9	0.05	100.13
S(~wetlands + lowslope)	3	105.0	4.7	0.035	98.95

Table 5. Parameter estimates, standard error (SE), and lower (LCI) and upper (UCI) confidence intervals for daily survivorship and overall survival; 29 days for nests (n = 37), and 60 days for fledglings (n = 18) and adult male (n = 20) and female (n = 11) rusty blackbird survival in northern New Hampshire from 2010–2012.

Survivorship Parameters	Survival Estimate	SE	LCI	UCI	Overall
Nest survival daily	0.975	0.26	0.95	0.99	0.48
Fledgling survivorship	0.988	0.336	0.97	0.99	0.49
Female survivorship	0.998	0.63	0.99	1	0.89 [1]
Male survivorship	0.996	0.44	0.99	1	0.79

[1] Female survivorship was estimated with a hypothetical record. See text for details.

Goodness-of-fit values for top habitat selection models were very close to or exceeded our threshold of 0.85 for nest sites (AUC = 0.87), fledglings (AUC = 0.87), and adults (AUC = 0.84). The Hosmer–Lemeshow test for nests (chi-squared = 9.0, p = 0.3) and fledglings (chi-squared = 15.8, p = 0.051) indicated no lack of fit. However, tests for adults indicated a difference between observed and expected proportions (chi squared = 28.1, $p \leq 0.001$). In the adult analysis, confidence intervals for estimates of two variables in the top model (softwood/mixed-wood sapling stands and low slope) overlapped one and thus were not useful for prediction; estimates for distance to streams in the nest analysis and for low slope in the fledgling analysis were similarly ambiguous (Table 3). The simplified adult model improved the chi-squared to 18.8, $p = 0.02$. Normal probability and residual plots indicated that all models satisfied assumptions of normality.

3.4. Blackbird Survival

Of the 18 fledglings, 20 adult males, and 11 adult females with transmitters in the survival analysis, 7 fledglings (39%), 5 males (25%), and 0 females succumbed to mortality during the life of their transmitter. Only one fledgling and one adult blackbird failed to survive the first week after fledging. There were no differences between years, drainages, or seasonal survival. Because the season variable was not important for explaining variation in survival of fledglings, including the period one week after fledging, males, females, and fledglings were analyzed together using age/sex dummy covariates; fledgling sex was unknown. The top model for blackbird survivorship included age and sex categories, with fledglings having the lowest survival rates (0.988, SE = 0.34), males with 0.996, SE = 0.44, and females with 0.998, SE = 0.63 (Table 5).

4. Discussion

4.1. Survival and Habitat Selection

Rusty blackbirds in our study selected (at a 30 m radius scale) a combination of habitat conditions for nesting and post-fledging, including shallow wetlands and low slope with softwood/mixed-wood sapling stands. This was the first study to assess the importance of distance to 1st to 6th order streams and found that this feature was selected for both nesting and post-fledging. Availability of streams in addition to wetlands may increase foraging opportunities because streams and wetlands have different availability of prey types throughout the season [31,49–52]. Early nestling development may be synchronized with the availability of small items typical of streams, such as Ephemeroptera, Plecoptera, and Trichoptera larvae, while dependent fledglings may require larger Odonata larvae, which are more abundant in wetlands (P. Wohner unpublished data). A diverse network of interconnected channels and impoundments, such as those created by beavers, may provide ideal foraging opportunities for rusty blackbirds [52].

We found that while habitat selection was similar between nest sites and post-fledging locations at the 30 m scale, the stages had one important difference: wetland cover. Likewise, in a study that included the same drainages, rusty blackbirds did not select wetlands for nesting at the 5 m scale [29]. At the 500 m scale, rusty blackbirds were reported more likely to select nesting habitat with increases in wetland cover in New Hampshire [29]. However, the confidence interval for the wetland estimate ranged from −2 to 13, overlapping 1, and is therefore inconclusive [29]. We concur that rusty blackbirds select nest sites independent of wetland cover at the 5, 30, or 500 m scale. However, wetlands are important to adults and fledglings at the 30 m scale. Concealment from predators may be more important than proximity to foraging areas during nesting and early post-fledging, while foraging opportunities quickly become a priority after fledging [9].

It is likely that proximity to wetlands is important for nest-site selection at a larger scale than has been studied thus far, i.e., >500 m [53]. Rusty blackbirds are highly mobile, traveling up to 2 km between wetlands and nesting areas, and have large home ranges (3.8–172.8 ha in Maine, equivalent to a circle with radius 35–741 m [54]). It is possible that cues for selection of nesting and foraging habitat are decoupled at different spatial scales, and rusty blackbirds select wetlands at a 1–5 km scale for foraging, and nesting habitat at a much smaller scale (0.05–0.5 km [53]). We propose that future nesting habitat selection studies that use multiple spatial scales including the 1–5 km scale would help to determine the density and size of wetlands important for rusty blackbirds at the home range scale and provide better predictive ability for rusty blackbirds [55].

Finding the appropriate scale could have important consequences for research and management [55]. For example, many surveys for nesting rusty blackbirds have been wetland-centric and typically only detect songbirds within 100–200 m of a point. In our study, 35 percent of nests were located >200 m from the nearest wetland. If rusty blackbird pairs forage at multiple wetlands and streams and nest in relatively distant uplands, wetland-based surveys likely underestimate occupancy and may result in biases [56]. The low detectability documented in other studies that only surveyed wetlands could be a consequence of survey design [28]. Other researchers have noted that point count-based surveys are not well suited for estimating the abundance or habitat use of breeding rusty blackbirds, even when broadcast calls are used [19].

Our nest success estimates (0.48) are well within the range reported by other rusty blackbird researchers (0.21–0.75; [27–29]) and are much higher than those reported for red-winged, yellow-headed, and Brewer's blackbirds (0.30–0.39; [30,57]). Regenerating clear-cuts were suspected ecological traps for nesting in rusty blackbirds in Maine [28]. However, we found evidence that regenerating clear-cuts support successful nesting by rusty blackbirds. Although we did not explicitly test stand origin, nest success did not decrease with increases in softwood sapling stands, 92% of which originated from even-aged forest management. This result is consistent with other research that found rusty blackbird nest success was independent of recent harvesting [26]. We agree with other researchers that nest success is unlikely to be limiting rusty blackbird populations [24,26].

As in other New England rusty blackbird breeding studies, where nest selection was assessed at 5 and 500 m radius scales [27,29], rusty blackbirds in our study selected young (5–15 years old) or stunted softwood/mixed wood stands at the 30 m radius scale for nesting. Many of the New England nesting stands were created by even-aged management and are composed of regenerating softwoods. For example, 75% of nests in Maine [28], 88% in Maine and New Hampshire [29], and 92% in this study were located in young, even-aged softwood stands. While rusty blackbirds seem to have an affinity for regenerating clear-cuts at the 5, 30, and 500 m scales, there may be a threshold for increasing proportions of regenerating softwood stands at larger scales than 500 m. Blackpoll warblers (*Setophaga striata*) were found to be positively associated with large proportions of clear-cut at the 115 m scale [6]. However, at the 1250 m scale, they were positively associated with clear-cuts only when < 5% of cover was clear-cut, above which the relationship was negative [6]. Thus, while even-aged forest management may be beneficial for rusty blackbirds at smaller scales, e.g., 5, 30, and 500 m radius, researchers should consider a scale larger than 500 m for future studies.

We observed low mortality rates for rusty blackbird fledglings in the first week after fledging (3%) compared to those reported for other songbirds, e.g., 21–81% for ovenbird [2], 11% for worm-eating warbler [8], and 90% for rose-breasted grosbeak (*Pheucticus ludovicianus*) fledglings [7]. All observed fledgling mortality took place in the first four days of a yellow warbler (*Setophaga petechi*) study [58]. Our high rusty blackbird fledgling survival contrasts with virtually every other songbird post-fledging study. We suggest that dense regenerating softwood stands approximately 5–15 years post-harvest, may afford < 1-week-old fledglings protection from potential predators while they are most vulnerable. Fledglings of songbirds that nest in mature hardwood stands with relatively open understories have sparse protective cover before moving to early successional habitats with dense vegetation [1,2,6,12]. Our overall fledgling survival during the 60-day post-fledging period was also relatively high (0.49) and is likely not an overwhelming factor contributing to population decline. However, we did not study survivorship after fledglings were completely independent from adults, nor during migration when young birds could experience substantial mortality [9]. Populations of fledgling barn swallows (*Hirundo rustica erythrogaster*) were found to be limited by the pre-migration phase [9].

While few studies have estimated adult survival during either nesting or post-fledging, adult barn swallow survival over 60 days was 0.92, SE = 0.11 for males and females together [59], compared to our estimates of 0.89 for females and 0.79 for males over the same time frame. Rusty blackbird male survival may truly be relatively low during the post-fledging period. However, the confidence intervals around our estimates are wide (0.54–1), likely due to small sample sizes. Transmitter harnesses may reduce survival, resulting in a survival estimate that is lower than actual survival in the general population. We regularly observed rusty blackbirds picking at their harnesses which could distract birds and expose them to higher predation. We have also recaptured blackbirds with transmitters embedded in their skin. Transmitter harnesses have been found to affect survival in sensitive species such as pileated woodpeckers (*Dryocopus pileatus*; [60]). Why harnesses would have a disproportionate effect on males over females is unknown but could be due to morphological differences between the sexes. High mortality of adult rusty blackbird males during the breeding season seems unusual, but if true, could itself be a key factor in population decline. Male survival during the breeding season warrants future study.

4.2. Recommendations

Nest success and adult and fledgling survival were not affected by any of the habitat variables we analyzed, which suggests that something other than breeding ground habitat may be limiting rusty blackbird populations. Thus, we expect that current forest management practices continue to create suitable landscapes for successful nesting and post-breeding survival. Our study adds support to a study in Nova Scotia, that found rusty blackbird habitat remains relatively abundant and well-distributed and is often located in wet lowlands which is a climate-resilient topographic landform [61]. Targeted habitat management for rusty blackbirds is likely to be unnecessary in many areas due to the species' use of regenerating softwood stands that are created by a variety of harvesting practices. We do recommend that harvest plans ensure the availability of at least one softwood stand 5 to 15 years post-harvest within 300 m of streams and shallow wetlands over time. In the Acadian forest, on ownerships where wildlife habitat is the focus of forest management, prioritizing harvests on poorly drained sites where trees grow more slowly could provide rusty blackbird nesting habitat for longer periods of time (≥40 years) [62].

To aid in finding high priority areas for research or conservation, we recommend overlaying the regression equation from our top models for nests and fledglings in the raster calculator in ARCMAP (i.e., for nests: −0.87−5.2 (distance to streams) + 0.01 (prop slope) + 0.02 (prop softwood/mixed-wood sapling stands). The nests and fledgling models had high goodness of fit, and together, could identify rusty blackbird habitat at the landscape level. We expect our regression equations will be applicable in the southeastern portion of the rusty blackbird's range in Acadian Forest, i.e., New Brunswick, Nova Scotia, northern New England, and the Adirondacks.

Finally, we hope that rusty blackbird researchers will move from traditional survey protocols [63,64] like presence-absence surveys to protocols such as those used for western yellow-billed cuckoos (*Coccyzus americanus occidentalis*), which also have large home ranges (16–91 ha) [65,66]. These surveys use broadcast calls every 100 m along transects in appropriate habitat [67,68]. (Spring surveys to locate pairs prior to nest searching for intensive research should avoid using broadcast calls, however). Future rusty blackbird research on the breeding ground should incorporate a landscape perspective and include multiple habitat types, including but not limited to dense young softwood stands, streams, seepages, and wetlands. Studying the species at a much larger scale than previously (e.g., >500m), may shed new light on wetland requirements during nesting.

Author Contributions: Conceptualization, C.R.F. and P.J.W.; methodology, C.R.F. and P.J.W.; formal analysis, P.J.W. and R.J.C.; investigation, C.R.F. and P.J.W.; resources, C.R.F. and R.J.C.; data curation, P.J.W. and C.R.F.; writing—original draft preparation, P.J.W.; writing—review and editing, P.J.W., C.R.F., R.J.C.; visualization, C.R.F. and P.J.W.; supervision, C.R.F.; project administration, C.R.F. and R.J.C.; funding acquisition, C.R.F., R.J.C., P.J.W. All authors have read and agreed to the published version of the manuscript.

Funding: This research was funded by The US Fish and Wildlife Service, New Hampshire Audubon, The Eastern Bird Banding Association, The Smithsonian Institution, The Natural Sciences and Engineering Research Council of Canada, and The University of Georgia.

Acknowledgments: A special thanks to all field assistants who spent considerable time in the clear-cuts and wetlands of New Hampshire from 2010–2012. They include S. Hribal, E. Prohl, R. Rabinovitz, H. Batcheller, E. Dancer, C. Ross, and J. Cosentino. Other supporters include those who granted access to rusty blackbirds on their lands including Umbagog National Wildlife Refuge, Wagner Forest Management, Conner, and Plum Creek. S. Edmonds and L. Powell contributed field expertise, on-the-ground assistance, and advice. Members of the International Rusty Blackbird Technical Working Group including L. Powell, S. Edmonds, S. Matsuoka, and D. Tessler provided technical advice on rusty blackbird surveying, nest searching, and capture. V. Jones from New Hampshire Audubon and N. Nibbelink from the University of Georgia assisted with GIS spatial analysis.

Conflicts of Interest: The authors declare no conflict of interest. The funders had no role in the design of the study; in the collection, analyses, or interpretation of data; in the writing of the manuscript, or in the decision to publish the results.

Appendix A

Summary of rusty blackbirds tagged by year and drainage in northern New Hampshire, USA from 2010–2012; includes total birds and number of points collected, and the subset of blackbirds and telemetry points retained in the analysis by adults (M = male and F = female) and fledglings. SWDI = Swift Diamond, MOLL = Mollidgewock, and INTE = Interior.

Year	Drainage Tagged	Total Birds	Total Points	Birds in Analysis	Adults in Analysis	Fledges in Analysis	Points in Analysis
2010	SWDI	8	167	4	1M,1F	2	64
2010	MOLL	8	130	6	3M,2F	1	73
2010	2010 Total	16	297	10	7	3	137
2011	SWDI	14	687	10	5M,2F	3	230
2011	MOLL	6	120	4	2M	2	60
2011	INTE	4	186	4	2M,1F	1	84
2011	2011 Total	24	993	18	12	6	374
2012	SWDI	12	486	10	5M,4F	1	211
2012	MOLL	0	0	0	0	0	0
2012	INTE	4	89	2	2F	0	49
2012	2012 Total	16	575	12	11	1	300
Total		56	1865	40	331	10	771

References

1. Vitz, A.C.; Rodewald, A.D. Can regenerating clearcuts benefit mature-forest songbirds? An examination of post-breeding ecology. *Biol. Conserv.* **2006**, *127*, 477–486. [CrossRef]
2. King, D.I.; Degraaf, R.M.; Smith, M.L.; Buonaccorsi, J.P. Habitat selection and habitat-specific survival of fledgling ovenbirds (*Seiurus aurocapilla*). *J. Zool.* **2006**, *269*, 414–421. [CrossRef]
3. Rush, S.A.; Stutchbury, B.J.M. Survival of fledgling hooded warblers (*Wilsonia citriana*) in small and large forest fragments. *Auk* **2008**, *125*, 183–191. [CrossRef]
4. Temple, S.; Cary, J. Modeling the dynamics of forest-interior bird populations in a fragmented landscape. *Conserv. Biol.* **1988**, *2*, 305–312. [CrossRef]
5. Faaborg, J.; Holmes, R.T.; Anders, A.D.; Bildstein, K.L.; Dugger, K.M.; Gauthreaux, S.A.; Heglund, P.J.; Hobson, K.A.; Jahn, A.E.; Johnson, D.H.; et al. Recent advances in understanding migration systems of New World land birds. *Ecol. Monogr.* **2010**, *80*, 3–48. [CrossRef]
6. Mitchell, G.W.; Taylor, P.D.; Warkentin, I.G. Assessing the function of broad-scale movements made by juvenile songbirds prior to migration. *Condor* **2010**, *112*, 644–654. [CrossRef]
7. Moore, L.C.; Stutchbury, B.J.M.; Burke, D.M.; Elliott, K. Effects of forest management on post-fledging survival of rose-breasted grosbeaks (*Pheucticus ludovicianus*). *Auk* **2010**, *127*, 185–194. [CrossRef]
8. Vitz, A.C.; Rodewald, A.D. Influence of condition and habitat use on survival of post-fledging songbirds. *Condor* **2011**, *113*, 400–411. [CrossRef]
9. Evans, D.; Hobson, K.A.; Kusack, J.W.; Cadman, M.D.; Falconer, C.M.; Mitchell, G.W. Individual condition, but not fledging phenology, carries over to affect post-fledging survival in a Neotropical migratory songbird. *IBIS* **2019**, *162*, 331–344. [CrossRef]

10. Anders, A.D.; Dearborn, D.C.; Faaborg, J.; Thompson, F.R., III. Juvenile survival in a population of neotropical migrant birds. *Conserv. Biol.* **1997**, *11*, 698–707. [CrossRef]

11. Burke, A.D.; Thompson, F.R., III; Faaborg, J. Variation in early-successional habitat use among independent juvenile forest breeding birds. *Wilson J. Ornithol.* **2017**, *129*, 235–246. [CrossRef]

12. Marshall, M.R.; DeCecco, J.A.; Williams, A.B.; Gale, G.A.; Cooper, R.J. Use of regenerating clearcuts by late-successional bird species and their young during the post-fledging period. *For. Ecol. Manag.* **2003**, *183*, 127–135. [CrossRef]

13. White, J.D.; Gardali, T.; Thompson, F.R., III; Faaborg, J. Resource selection by juvenile Swainson's thrushes during the post-fledging period. *Condor* **2005**, *107*, 388–401. [CrossRef]

14. Jenkins, J.M.A.; Thompson, F.R., III.; Faaborg, J. Species-specific variation in nesting and post-fledging resource selection for two forest breeding migrant songbirds. *PLoS ONE* **2017**, *12*, e0179524. [CrossRef] [PubMed]

15. Delancey, C.D.; Islam, K. Post-fledging habitat use in a declining songbird. *PeerJ* **2019**, *7*, e7358. [CrossRef] [PubMed]

16. Thompson, B.C.; Knadle, G.E.; Brubaker, D.L.; Brubaker, K.S. Nest success is not an adequate comparative estimate of avian reproduction. *J. Field Ornithol.* **2001**, *72*, 527–536. [CrossRef]

17. Avery, M.L. Rusty Blackbird (*Euphagus carolinus*). In *The Birds of North America*, version 2.0.; Rodewald, P.G., Ed.; Cornell Lab of Ornithology: Ithaca, NY, USA, 2013. [CrossRef]

18. Glennon, M.J. Dynamics of boreal birds at the edge of their range in the Adirondack Park, NY. *Northeast. Nat.* **2014**, *21*. [CrossRef]

19. McNulty, S.; Glennon, M.; McCormack, M. Rusty blackbirds in New York State: Ecology, current status, and future. *Adirond. J. Environ. Stud.* **2015**, *20*, 87–99.

20. Greenberg, R.; Droege, S. On the decline of the rusty blackbird and the use of ornithological literature to document long-term population trends. *Conserv. Biol.* **1999**, *13*, 553–559. [CrossRef]

21. Niven, D.K.; Sauer, J.R.; Butcher, G.S.; Link, W.A. Christmas Bird Count provides insights into population change in land birds that breed in the boreal forest. *Am. Birds* **2004**, *58*, 10–20.

22. Sauer, J.R.; Hines, J.E.; Fallon, J. *The North American Breeding Bird Survey, Results and Analysis 1966–2005*; Version 6.2.; USGS Patuxent Wildlife Research Center: Laurel, ML, USA, 2006.

23. Greenberg, R.; Matsuoka, S.M. Rusty blackbird: Mysteries of a species in decline. *Condor* **2010**, *112*, 770–777. [CrossRef]

24. Greenberg, R.; Demarest, D.W.; Matsuoka, S.M.; Mettke-Hofmann, C.; Evers, D.; Hamel, P.B.; Luscier, J.; Powell, L.L.; Shaw, D.; Avery, M.L.; et al. Understanding declines in rusty blackbirds. In *Boreal Birds of North America: A Hemispheric View of Their Conservation Links and Significance*; Studies in Avian Biology: 41; Wells, J.V., Ed.; University of California Press: Berkeley, CA, USA, 2011; pp. 107–126.

25. U.S. Fish and Wildlife Service. *Birds of Conservation Concern*; U.S. Fish and Wildlife Service, Division of Migratory Bird Management: Arlington, VA, USA, 2008.

26. International Union for Conservation of Nature and Natural Resources [IUCN]. *Euphagus Carolinus*, IUCN Red List of Threatened Species, version 2010. 2; International Union for Conservation of Nature and Natural Resources: Gland, Switzerland, 2010.

27. Matsuoka, S.; Shaw, D.; Sinclair, P.H.; Johnson, J.A.; Corcoran, R.M.; Dau, N.C.; Meyers, P.M.; Rojek, N.A. Nesting ecology of the rusty blackbird in Alaska and Canada. *Condor* **2010**, *112*, 810–824. [CrossRef]

28. Powell, L.L.; Hodgman, T.P.; Glanz, W.E.; Osenton, J.D.; Fisher, C.M. Nest-site selection and nest survival of the rusty blackbird: Does timber management adjacent to wetlands create ecological traps? *Condor* **2010**, *112*, 800–809. [CrossRef]

29. Buckley Luepold, S.H.; Hodgman, T.P.; McNulty, S.A.; Cohen, J.; Foss, C.R. Habitat selection, nest survival, and nest predators of rusty blackbirds in northern New England, USA. *Condor* **2015**, *117*, 609–623. [CrossRef]

30. Martin, T.E. Avian life history evolution in relation to nest sites, nest predation, and food. *Ecol. Monogr.* **1995**, *65*, 101–127. [CrossRef]

31. Pachomski, A.L.; McNulty, S.; Foss, C.R.; Cohen, J.; Farrell, S. Rusty Blackbird (*Euphagus carolinus*) foraging habitat and prey availability in New England: Implications for conservation of a declining boreal bird species. *Diversity* **2020**, *12*.

32. Schmidt, K.A.; Rush, S.A.; Ostfeld, R.S. Wood thrush nest success and post-fledging survival across a temporal pulse of small mammal abundance in an oak forest. *J. Anim. Ecol.* **2008**, *77*, 830–837. [CrossRef]

33. Thomas, D.W.; Blondel, J.; Perret, P.; Lambrechts, M.M.; Speakman, J.R. Energetic and fitness costs of mismatching resource supply and demand in seasonally breeding birds. *Science* **2001**, *291*, 2598–2600. [CrossRef]

34. Bailey, R.G. *Description of the Ecoregions of the United States*; 2nd edition revised and expanded. Misc. Publ. No. 1391 (rev.); USDA Forest Service: Washington, DC, USA, 1995.

35. Appalachian Mountain Club. *Ecological Atlas of the Upper Androscoggin River Watershed*; Appalachian Mountain Club: Boston, MA, USA, 2003.

36. U.S. Climate Data. Available online: https://www.usclimatedata.com/climate/berlin/new-hampshire/united-states/usnh0020 (accessed on 14 May 2020).

37. Rappole, J.H.; Tipton, A.R. New harness design for attachment of radio transmitters to small passerines. *J. Field Ornithol.* **1991**, *62*, 335–337.

38. Lele, S.R.; Merrill, E.H.; Keim, J.; Boyce, M.S. Selection, use, choice and occupancy: Clarifying concepts in resource selection studies. *J. Anim. Ecol.* **2013**, *82*, 1183–1191. [CrossRef]

39. R Core Team. *R: A Language and Environment for Statistical Computing*; R Foundation for Statistical Computing: Vienna, Austria, 2012.

40. Fieberg, J.; Matthiopoulos, J.; Hebblewhite, M.; Boyce, M.S.; Frair, J.L. Correlation and studies of habitat selection: Problem, red herring or opportunity? *Philos. Trans. R. Soc. B* **2010**, *365*, 2233–2244. [CrossRef] [PubMed]

41. Neter, J.; Wasserman, W.; Kutner, M.H. *Applied Linear Statistical Models R.D.*; Irwin: Homewood, IL, USA, 1990.

42. Burnham, K.P.; Anderson, D.R. *Model. Selection and Inference: An Information-Theoretic Approach*; Springer: New York, NY, USA, 2002.

43. Royall, R.M. *Statistical Evidence: A Likelihood Paradigm*; Chapman and Hall: London, UK, 1997.

44. Hosmer, D.W.; Lemeshow, S. *Applied Logistic Regression*, 1st ed.; John Wiley and Sons: New York, NY, USA, 1989.

45. Ott, R.L.; Longnecker, M. *An Introduction to Statistical Methods and Data Analysis*, 1st ed.; Wadsworth Group: Pacific Groove, CA, USA, 2001.

46. Laake, J.L. *RMark: An R Interface for Analysis of Capture-Recapture Data with MARK*; U.S. Department of Commerce: Seattle, WA, USA, 2017.

47. Cooch, E.; White, G. *Program. MARK: A Gentle Introduction*, 20th ed. 2018. Available online: http://www.phidot.org/software/mark/docs/book/ (accessed on 1 June 2020).

48. White, G.C.; Burnham, K.P. Program. MARK: Survival estimation from populations of marked animals. *Bird Study* **1999**, *46*, S120–S139.

49. Batzer, D.P.; Rader, R.B.; Wissinger, S.A. *Invertebrates in Freshwater Wetlands of North America: Ecology and Management*; John Wiley and Sons: New York, NY, USA, 1999.

50. Arndt, A.; Domdei, J. Influence of beaver ponds on the macroinvertebrate benthic community in lowland brooks. *Pol. J. Ecol.* **2011**, *59*, 799–811.

51. McClure, C.J.W.; Rolek, B.W.; McDonald, K.; Hill, G.E. Climate change and the decline of a once common bird. *Ecol. Evol.* **2012**, *2*, 370–378. [CrossRef] [PubMed]

52. Bush, B.M.; Wissinger, S.A. Invertebrates in beaver-created wetlands and ponds. In *Invertebrates in Freshwater Wetlands*; Batzer, D., Boix, D., Eds.; Springer International Publishing: Cham, Switzerland, 2016; pp. 411–449.

53. Westwood, A.R. Conservation of Three Forest Landbird Species at Risk: Characterizing and Modelling Habitat at Multiple Scales to Guide Management Planning. Ph.D. Thesis, Dalhousie University, Halifax, NS, Canada, 2016.

54. Powell, L.L.; Hodgman, T.P.; Glanz, W.E. Home ranges of Rusty Blackbirds breeding in wetlands: How much would buffers from timber harvest protect habitat? *Condor* **2010**, *112*, 834–840. [CrossRef]

55. Bohnett, E.; Hulse, D.; Ahmad, B.; Hoctor, T. Multi-level, multi-scale modeling and predictive mapping for jaguars in the Brazilian Pantanal. *Open J. Ecol.* **2020**, *10*, 243–263. [CrossRef]

56. Schaefer, J.A.; Mayor, S.J. Geostatistics reveal the scale of habitat selection. *Ecol. Model.* **2007**, *209*, 401–406. [CrossRef]

57. Duquette, C.A.; Hovick, T.J.; Limb, R.F.; McGranahan, D.A.; Sedivec, K.K. Restored fire and grazing regimes influence nest selection and survival in Brewer's Blackbirds Euphagus Cyanocephalus. *Acta Ornithol.* **2019**, *54*, 171–180. [CrossRef]

58. Hepp, M.; Ware, L.; van Oort, H.; Beauchesne, S.M.; Cooper, J.M.; Green, D.J. Postfledging survival and local recruitment of a riparian songbird in habitat influenced by reservoir operations. *Avian Conserv. Ecol.* **2018**, *13*, 12. [CrossRef]

59. Gruebler, M.E.; Korner-Nievergelt, F.; Naef-Daenzer, B. Equal nonbreeding period survival in adults and juveniles of a long-distant migrant bird. *Ecol. Evol.* **2014**, *4*, 756–765. [CrossRef]

60. Noel, B.L.; Bednarz, J.C.; Ruder, M.G.; Keel, M.K. Effects of radio-transmitter methods on pileated woodpeckers: An improved technique for large woodpeckers. *Southeast. Nat.* **2013**, *12*, 399–412. [CrossRef]

61. Bale, S.; Beazley, K.F.; Westwood, A.; Bush, P. The benefits of using topographic features to predict climate-resilient habitat for migratory forest landbirds: An example for the Rusty Blackbird, Olive-sided Flycatcher, and Canada Warbler. *Condor* **2020**, *122*. [CrossRef]

62. Burns, R.M.; Honkala, B.H. Silvics of North America: Volume 1. Conifers. In *Agriculture Handbook 654*; U.S. Forest Service, Department of Agriculture: Washington, DC, USA, 1990.

63. Powell, L.L. *Long Term Monitoring Plan for Rusty Blackbirds in the Atlantic Northern Forest*; Version 1.8; University of Maine: Orono, ME, USA, 2008.

64. Powell, L.L.; Hodgman, T.P.; Fiske, I.J.; Glanz, W.E. Habitat occupancy of rusty blackbirds (*Euphagus carolinus*) breeding in northern New England, USA. *Condor* **2014**, *116*, 122–133. [CrossRef]

65. Sechrist, J.; Ahlers, D.D.; Zehfuss, K.P.; Doster, R.H.; Paxton, E.H.; Ryan, V.M. Home range and use of habitat of western yellow-billed cuckoos on the middle Rio Grande, New Mexico. *Southwest. Nat.* **2013**, *58*, 411–419. [CrossRef]

66. Halterman, M.D. Sexual Dimorphism, Detection Probability, Home Range, and Parental Care in the Yellow-Billed Cuckoo. Ph.D. Thesis, University of Nevada, Reno, NV, USA, 2009.

67. Laymon, S.A.; Halterman, M.D. A Proposed Habitat Management Plan for Yellow-Billed Cuckoos in California. In *Proceedings of the California Riparian Systems Conference: Protection, Management, and Restoration for the 1990s, Davis, CA, USA, 22–24 September 1988*; Gen. Tech. Rep. PSW-GTR-110; Abell, D.L., Ed.; Pacific Southwest Forest and Range Experiment Station, Forest Service, U.S. Department of Agriculture: Berkeley, CA, USA, 1989; pp. 272–277.

68. Halterman, M.D.; Johnson, M.J.; Holmes, J.A.; Laymon, S.A. *A Natural History Summary and Survey Protocol for the Western Distinct Population Segment of the Yellow-Billed Cuckoo*; U.S. Fish and Wildlife Service: Washington, DC, USA, 2015; p. 45.

 *diversity*

Article

Flock Size Predicts Niche Breadth and Focal Wintering Regions for a Rapidly Declining Boreal-Breeding Passerine, the Rusty Blackbird

Brian S. Evans [1,2], **Luke L. Powell** [1,3,4,*], **Dean W. Demarest** [5], **Sinéad M. Borchert** [6] and **Russell S. Greenberg** [1,†]

[1] Migratory Bird Center, Smithsonian Conservation Biology Institute, National Zoological Park, P.O. Box 37012, Washington, DC 20013, USA; evansbr@si.edu (B.S.E.)
[2] Department of Biology, Georgetown University, 37th and O Streets NW, Washington, DC 20057, USA
[3] Institute of Animal Health and Comparative Medicine, Graham Kerr Building, University of Glasgow, Glasgow G128QQ, UK
[4] Department of Biosciences, Durham University, Stockton Road, Durham DH13LE, UK
[5] U.S. Fish & Wildlife Service, Division of Migratory Birds, Atlanta, GA 30345, USA; Dean_Demarest@fws.gov
[6] U.S. Fish & Wildlife Service, Panama City Field Office, Panama City, FL 32405, USA; sinead_borchert@fws.gov
* Correspondence: Luke.L.Powell@gmail.com
† Deceased.

Abstract: Once exceptionally abundant, the Rusty Blackbird (*Euphagus carolinus*) has declined precipitously over at least the last century. The species breeds across the Boreal forest, where it is so thinly distributed across such remote areas that it is extremely challenging to monitor or research, hindering informed conservation. As such, we employed a targeted citizen science effort on the species' wintering grounds in the more (human) populated southeast United States: the Rusty Blackbird Winter Blitz. Using a MaxEnt machine learning framework, we modeled patterns of occurrence of small, medium, and large flocks (<20, 20–99, and >99 individuals, respectively) in environmental space using both Blitz and eBird data. Our primary objective was to determine environmental variables that best predict Rusty Blackbird occurrence, with emphasis on (1) examining differences in key environmental predictors across flock sizes, (2) testing whether environmental niche breadth decreased with flock size, and (3) identifying regions with higher predicted occurrence (hotspots). The distribution of flocks varied across environmental predictors, with average minimum temperature (~2 °C for medium and large flocks) and proportional coverage of floodplain forest having the largest influence on occurrence. Environmental niche breadth decreased with increasing flock size, suggesting an increasingly restrictive range of environmental conditions capable of supporting larger flocks. We identified large hotspots in floodplain forests in the Lower Mississippi Alluvial Valley, the South Atlantic Coastal Plain, and the Black Belt Prairie.

Keywords: Black Belt Prairie; citizen science; conservation; machine learning; niche modeling; group size; habitat use; species distribution models

Citation: Evans, B.S.; Powell, L.L.; Demarest, D.W.; Borchert, S.M.; Greenberg, R.S. Flock Size Predicts Niche Breadth and Focal Wintering Regions for a Rapidly Declining Boreal-Breeding Passerine, the Rusty Blackbird. *Diversity* **2021**, *13*, 62. https://doi.org/10.3390/d13020062

Academic Editor: Michael Wink
Received: 30 November 2020
Accepted: 29 January 2021
Published: 4 February 2021

Publisher's Note: MDPI stays neutral with regard to jurisdictional claims in published maps and institutional affiliations.

1. Introduction

The Rusty Blackbird (*Euphagus carolinus*) is a Nearctic icterid that exhibits near exclusive reliance on forested wetland habitats in boreal, transitional and deciduous biomes throughout its annual cycle [1–3]. The species is among the most steadily and precipitously declining temperate North American landbirds, [4–6] and is recognized as "Vulnerable" by the International Union for the Conservation of Nature. There is some anecdotal evidence that factors in boreal breeding regions of northeast North America may be limiting the population (timber management and nest success: [7]; mercury: [8]). However, demographic rates for Rusty Blackbirds across the Boreal are similar to those of other songbirds [1]), suggesting that the overall population is likely limited by factors outside the breeding grounds.

Specifically, habitat modification on the species' wintering grounds has been implicated as a likely driver of the decline [5,9], stressing the importance of studying this period of the species' life cycle. Further, population monitoring and broad assessments of habitat selection across the boreal breeding range are extremely challenging given the general remoteness and the very low density of roads, humans, and Rusty Blackbirds themselves. The Breeding Bird Survey, for example, detects so few Rusty Blackbirds that with the exception of more developed parts of Alaska it is not possible to use these data to monitor population trends. All of these factors coupled with the species' considerably more social (flocking) nature on the wintering grounds and the high densities of citizen scientists (birders) in the southeast United States point to the advantages of studying the species during winter. The wintering grounds are not without their challenges, however; diffuse and localized distribution, nomadism, and difficulty in accessing often ephemeral wetland habitats constrain our ability to dependably study wintering Rusty Blackbirds [9]. Conventional means of investigation, such as use of mark-recapture (e.g., to evaluate survival as a function of winter habitat metrics), may not be viable due to high vagility and low site fidelity between years (e.g., [10,11].

Once reported in wintering flocks numbering tens of thousands [4], Rusty Blackbirds are now occasionally observed in flocks approaching a few thousand individuals, with much smaller groups being more common [12,13]. Still, flocking behavior in Rusty Blackbirds may provide a proximate cue allowing researchers to identify and characterize winter habitat (e.g., key sites, condition, relative quality or suitability for foraging). Foraging theory predicts that abundance and distribution of food resources influence group size, with more resource rich environments favorable to flocks of larger size (sensu [14–18]). Relationships between group size and resource availability have been shown in a wide variety of taxa, including primates (e.g., *Lemur catta* [19]), fish (e.g., *Fundulus diaphanous* [20]), and birds (e.g., *Junco phaeonotus* [21]). Where food resources are less available, interference competition can moderate flock size as individuals seek to optimize foraging efficiency beyond the confines of competitive groups [15]. Demographic studies in a number of bird species support that the size of conspecific aggregations is positively associated with higher quality habitats [22,23].

Because flock size is expected to vary as a function of habitat quality or compatibility, evaluating environmental features associated with flocks of different sizes may provide evidence for conditions and locations most favorable for greater numbers of wintering Rusty Blackbirds. Assessing distributions in environmental space, as opposed to geographic space, is necessary for the species as wintering individuals often travel long distances across the landscape, and thus site level habitat characteristics have not consistently predicted occupancy [2]. An understanding of such relationships could offer immediate relevance in informing conservation programs and facilitating further research and monitoring supporting full annual cycle stewardship of this species, as well as yield new insights regarding niche dynamics in flocks of different sizes.

Here, we evaluate the relationships of climatic and landcover attributes in predicting the occurrence of Rusty Blackbirds across their wintering range and ask whether these relationships vary with flock size. Previous studies have observed that, at local scales, large flocks of Rusty Blackbirds are constrained to a narrow set of environmental conditions and geographic areas, whereas smaller flocks are comparatively widespread (e.g., [12,24,25]). We sought to quantify these observations at the spatial extent of winter range of the species, predicting that (1) large flocks of Rusty Blackbirds (>100 observed individuals) occupy a different environmental niche than small or medium flocks (<20, 20–99, respectively) and (2) large flocks occupy a narrower environmental niche breadth than small or medium flocks [26]. We define niche breadth as the multidimensional range of specific biotic and abiotic conditions that a group of organisms occupies [27–29]. We used occurrence data from two citizen science programs, the Rusty Blackbird Winter Blitz (Blitz; http://rustyblackbird.org/outreach/migration-blitz/) and eBird [30], to test these predictions by modeling the distributions of Rusty Blackbird flock size classes in environmental space.

In a closely related project, we performed a three-year "Spring Blitz" for migrating Rusty Blackbirds, but those results are addressed elsewhere [31]. We compared biotic and abiotic attributes that best predicted occurrences of flock size classes and evaluated differences in environmental niche breadth among classes. We suggest focal regions for wintering Rusty Blackbirds (hotspots), defined as regions with higher predicted occurrence, and discuss implications to future research and conservation. Finally, given the proliferation of citizen science monitoring projects, we evaluated whether the Blitz, a targeted volunteer effort requiring considerable investment and coordination, yielded observational data that improved predictive strength of relationships relative to use of comparable eBird data alone.

2. Methods

We used a maximum entropy (MaxEnt) modeling approach to examine habitat suitability across flock size classes and observational methods. In species distribution modeling, habitat suitability is defined as the likelihood of occurrence of a species in association with environmental variables ([32], but see [33]). MaxEnt is a machine-learning method that compares occurrence data with that of background samples in environmental space (see Supplementary File S1 1.1; [34]). In this context, training samples represent subsets of locations in which Rusty Blackbirds were observed. Background samples represent subsets of locations that were sampled but no Rusty Blackbirds were observed.

We used Rusty Blackbird observations submitted from the Blitz, as well as those collected through traditional eBird [30] protocols. The Blitz was a coordinated effort among citizen scientists to search for Rusty Blackbirds across the species' wintering range by conducting traveling counts of 8 km or less during target dates of 1 January–28 February 2009–2011. Blitz observations were submitted using a special eBird portal (https://ebird.org/home) and recorded date, location, species, numbers, effort and other attributes (see [35]). To supplement Blitz data, we used traditional eBird checklists that reported Rusty Blackbirds on traveling counts conducted in 2009–2011 during corresponding Blitz periods [30]. We considered each eBird and Blitz checklist as an independent observation, omitting duplicates (i.e., observations submitted by multiple persons). We georeferenced observation locations with a 4-km resolution raster grid (see below). To avoid pseudoreplication, if a grid cell contained >1 independent observation we retained only the observation recording the highest Rusty Blackbird count.

We classified Rusty Blackbird observations by flock size (small: <20; medium: 20–99; large: >99 individuals), and sampling protocol (Blitz or eBird). Flock size classes were determined by the frequency of observations of a given number of individuals and approximate the lower (small flocks) and upper (large flocks) quantiles of observations within the combined pool of eBird and Blitz samples. While we use the term "flock" to designate the number of birds observed, both Blitz and traditional eBird observations were reported as traveling counts. Thus, it is possible that in some cases, aggregates designated as "large flocks" may be comprised of several small or medium-sized true flocks.

Given that flocks aggregate in common roosts at night and may be observed traveling in smaller groups during the day, we examined whether the time of observation biased our results. We determined the time of observation for each checklist used in our analysis (i.e., the maximum observed Rusty Blackbirds within a given raster cell). All time values were standardized to Coordinated Universal Time as a function of their geographic coordinates. Given that day length varies by geographic location and over the two-month period of this study, we used the R package *activity* [36] to transform the times of observation to solar times, which are defined here as radian time values relative to civil sunrise times for a given date and location [37,38]. We then fit Von Mises circular kernel density distributions to detections within each flock size class [36]. Following Ridout and Linkie [39], we calculated the degree of overlap ($\hat{\Delta}_4$) between fitted distributions and used randomization (with replacement; $n = 10{,}000$ iterations) to test the null hypothesis that detections come for the same distribution ($\alpha = 0.05$). For each of the flock size class pairs, we failed to reject the null hypothesis (File S1), providing supportive evidence that the time of day in which an observation occurred was unlikely to introduce bias into our results.

A limitation to occurrence-only modeling methods such as MaxEnt is that occurrence data, especially data collected opportunistically, are often biased toward areas of higher sampling effort (e.g., as a function of accessibility, convenience, or availability of human observers; see [40]. This confounds inferences regarding distribution, habitat use and environmental niche, and may cause inflated measures of model performance [41,42]. To mitigate this potential bias in our models, we generated background data by summarizing eBird checklists (2009–2011; n = 13,218 checklists) across each 4 km grid cell. In doing so, this method assumes that the sampling bias associated with checklists observing Rusty Blackbirds is similar to that of eBird checklists where Rusty Blackbirds were not observed [43,44]. By using random eBird sampling locations as background points, rather than random locations from the entire study extent, we were able to construct models that evaluate the occurrence of Rusty Blackbird observations in environmental space relative to points on the landscape presumed to be representative of any geographic biases in our Rusty Blackbird observations (see [45]). Additionally, we limited background point selection, and thus the extent of our distribution models [46], to the geographic extent that contained 99% of Rusty Blackbird observations, which corresponded to the land area of the Eastern United States, bounded in the west at a longitude of $-100°$.

We constructed models with up to 14 environmental covariates to predict Rusty Blackbird habitat suitability, including two climatic and 12 land cover covariates (see Table S1). We used the *raster* package in Program R [47,48] for all environmental layer processing. We obtained mean precipitation and minimum temperature data for the months of January and February 2009–2011 (4-km resolution, [49]). Though there was considerable annual variation in average minimum temperature and precipitation within our study area, small annual sample sizes of Rusty Blackbird observations necessitated averaging these covariates across the three winters. We obtained 30-m-resolution land cover data from the US Gap Analysis project [50] and reclassified Gap classes into the following 12 categories: floodplain forest, hardwood forest, plantation hardwood forest, upland forest, mixed forest, woodland, shrub, wetland, grassland, pasture, row crops, and developed land (Figure 1, Table S1). We calculated the mean proportional cover of each category within a 4 km resolution raster grid. This spatial scale is roughly equivalent to winter home range sizes observed for Rusty Blackbirds [12]. The values of all environmental variables were extracted to the spatial locations of Rusty Blackbird observations and eBird background samples.

Models were constructed using MaxEnt 3.3.3 [34] and implemented in the R package *dismo* [51]. We randomly partitioned observations into five replicates (k-fold partitioning with cross-validation), with 80% of observations for each replicate used to fit the MaxEnt model (i.e., training samples) and the remaining 20% of the observations held aside to evaluate model performance (i.e., test samples). To minimize the number of features used in model construction and maximize the interpretability of individual covariate effects, we used linear feature constraints of environmental covariates in model construction [52]. However, because Rusty Blackbirds may exhibit a non-linear distribution in response to minimum winter temperatures, we included a quadratic form of the minimum temperature model covariate. To limit model over-fitting, we calibrated models by selecting the most parsimonious models for each flock size class (see Supplementary File S1 1.2).

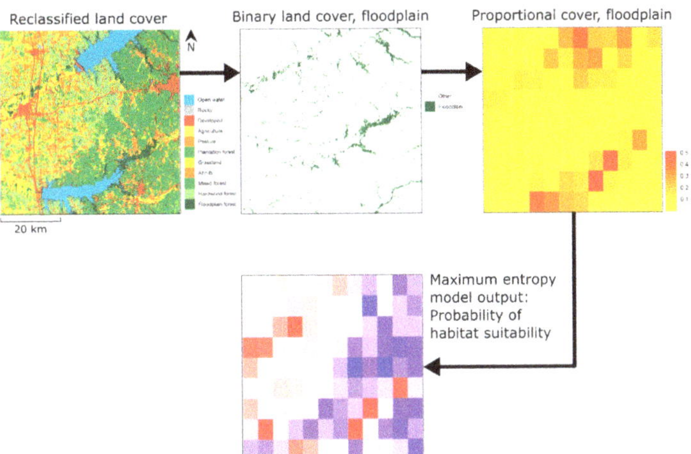

Figure 1. Processing steps from classified land cover layer to distribution modeling.

3. Model Evaluation

We evaluated model performance by comparing model sensitivity (the true positive rate—the proportion of correctly identified samples at a given threshold of habitat suitability) and specificity (the false positive rate—the proportional predicted area for the model) for each flock size class and observational method. To do so, we assessed the area under the receiver operator curve (AUC) for test samples. AUC describes the sensitivity and specificity of a model at a given threshold of suitability and represents how well the model predicts the Rusty Blackbird observations (see Figure 2). AUC values of 0.5 represent equivalent model performance relative to random, 0.5–0.6 "poor" performance, 0.6–0.7 "fair" performance, 0.7–0.8 "good" performance, 0.8–0.9 "very good" performance, and 0.9–1.0 "excellent" model performance [53,54]. We tested whether the predictive capacity of models varied by flock size by comparing training AUC across folds for a given flock size class against a null distribution developed by permuting two flock size classes (i.e., suitability models developed by shuffling large and small flock observations). To determine whether Blitz data improved model performance, we compared observed AUC against a null distribution developed by randomizing eBird and Blitz samples.

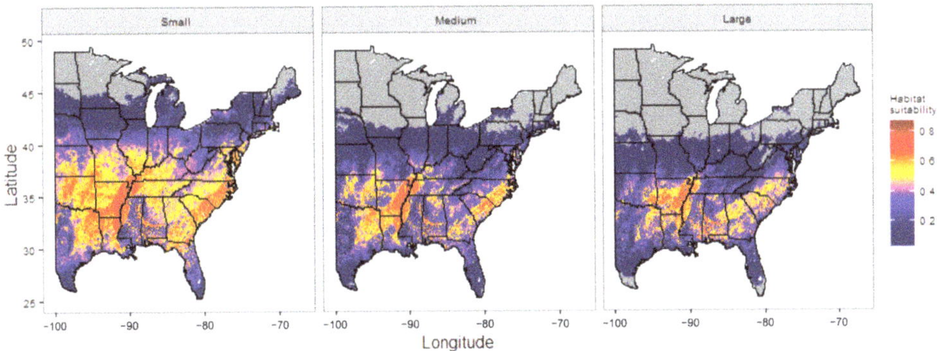

Figure 2. Habitat suitability estimates for small, medium, and large flocks of Rusty Blackbirds (<20, 20–99, and >99 individuals, respectively). Across all flock size classes, maps suggest focal wintering regions within the Lower Mississippi Alluvial Valley, Black Belt Prairie, and South Atlantic Coastal Plain. These are also available in Supplementary File S2 as spatially-referenced kmz files suitable for Google Earth.

4. Influence of Environmental Variables

To determine which environmental variables best predict Rusty Blackbird occurrence for each flock size class, we compared the influence of variable inclusion or removal on model performance. The coefficient for each variable (λ) describes the influence of that variable on suitability estimates, with positive values representing enhanced suitability and negative values representing lower suitability. To determine contribution of a variable to model fit, we evaluated jackknife estimates of the increase and decrease in AUC with the inclusion and removal of a given variable, averaged across replicate runs [55]. Additionally, we examined the difference in the occurrence of environmental covariates between flock sizes, observation protocols, and background data by permuting covariate values between classes. We compared empirical and null distributions of each environmental variable with a one-tailed test of the null hypothesis that the distributions are not statistically different ($\alpha = 0.05$) using the R package *permute* ($n = 10,000$, [56]).

5. Difference in Environmental Niche by Flock Size

To assess differences in Rusty Blackbird environmental niche breadth among flock size classes, we evaluated the predicted niche breadth using ENMTools [57]. This metric follows Levins' [29] definition of niche breadth, which describes the degree of uniformity of resource states within a distribution of individuals. Niche breadth is standardized to range between 0 and 1, where 0 represents a habitat specialist and 1 represents a habitat generalist. To evaluate our prediction that niche breadth will vary by flock size, we compared the empirical niche breadth metric for each size class with species distribution maps developed by permuting flock size class assignments. We further explored Rusty Blackbird use of niche space by assessing model prevalence, which is the MaxEnt estimate of the average probability of presence at background sites [58], and fractional predicted area of suitable habitat. We calculated fractional predicted area at the logistic threshold of equal model sensitivity and specificity. We statistically evaluated differences in prevalence and fractional predicted area between flock size classes by comparing the observed values for a given flock with a null distribution developed by permuting flock size class labels ($\alpha = 0.05$).

To test whether realized ecological niches differed among flock size classes, we evaluated the suitability distributions using niche equivalency analysis and assessed the distribution of samples across individual environmental variables. Niche equivalency between flock size classes was determined by calculating the degree of similarity (modified Hellinger distances, *I*) between habitat suitability maps [59] and implemented in the R package *phyloclim* [60]. *I* ranges in value from 0, in which there is no overlap in the environmental niche between classes, to 1, in which the two classes have identical distributions [59]. We compared observed similarity against a null distribution developed by randomly permuting flock class labels ($n = 99$ permutations). We assessed statistical significance ($\alpha = 0.05$) by comparing the observed and null distributions, with a one-tailed test of the null hypothesis that environmental niche models are equivalent. We compared the observed distributions of flock size classes of Rusty Blackbirds for each environmental variable using two-tail Kolmogorov–Smirnov tests of the null hypothesis that samples were drawn from the same distribution.

6. Results

The Blitz produced 678 independent checklists east of $-100°$ longitude. eBird provided an additional 1429 traditional checklists with Rusty Blackbird observations corresponding to the same area and time periods. Limiting checklists to one per 4-km raster grid cell resulted in 495 Blitz and 714 traditional eBird checklists (Table 1). Predicted suitability maps (Figure 2) across all flock size classes suggested hotspots within the Lower Mississippi Alluvial Valley (LMAV), Black Belt Prairie (BBP), and South Atlantic Coastal Plain (SACP). For small flocks, there were additional broad areas of moderately high to high predicted suitability, whereas for large flocks the extent of high predicted suitability was more clearly limited to the aforementioned areas albeit much more restricted within the LMAV.

Table 1. Number of checklists used in this analysis by flock size class and observation method. The sample size reflects the number of samples used after subsetting samples to one observation per raster grid cell.

Sampling Method	Flock Size Class			Total
	Small (1–19)	Medium (20–99)	Large (>99)	
Blitz	281	128	86	495
eBird	387	234	93	714
Total	668	362	179	1209

7. Model Performance and Environmental Correlates by Flock Size

Model performance increased with flock size, with fair performance for small flocks (AUC: 73.7 across samples) and good performance for medium and large flocks (AUCs 83.2 and 88.0 across samples, respectively; Figure 3B). Compared to models using eBird data alone, models augmented with Blitz data showed improved performance for medium and large flocks as suggested by higher AUC values (0.73 ± 0.01 and 0.72 ± 0.01 vs. 0.83 ± 0.02 and 0.88 ± 0.02, respectively; Figure 3B).

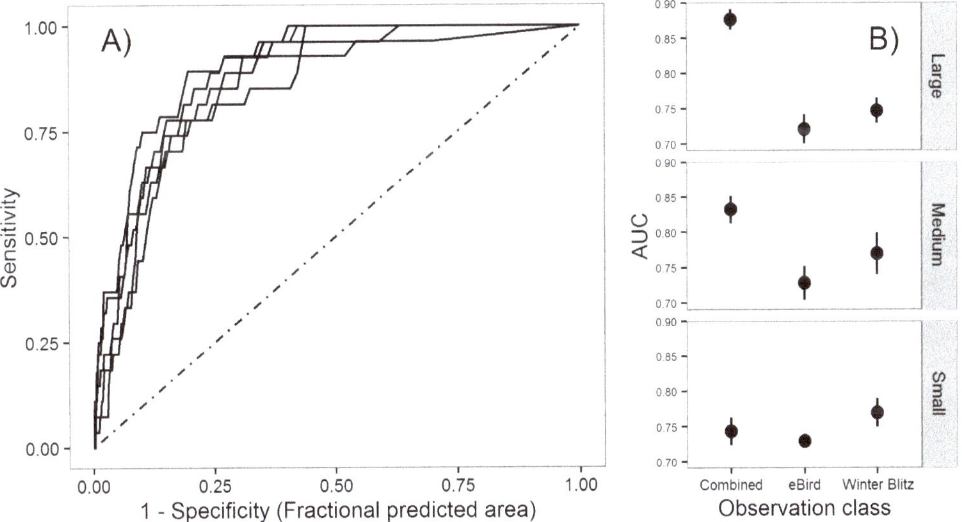

Figure 3. Receiver operator curves (**A**) and Area Under the curve (AUC; **B**) for small, medium, and large flocks of Rusty Blackbirds using Blitz data, eBird data alone, or Blitz and eBird data ("combined"). Solid lines (**A**) represent receiver operator curves for replicate runs across sampling methods and flock size classes. AUC values of 0.5 represent equivalent MaxEnt model performance relative to random, 0.5–0.6 "poor" performance, 0.6–0.7 "fair", 0.7–0.8 "good", 0.8–0.9 "very good", and 0.9–1.0 "excellent".

Among the environmental covariates, average minimum temperature most strongly predicted habitat suitability, contributing >60% of the models' predictive capacity for each flock size class (Figures 4 and 5A). Larger flock sizes were associated with higher average minimum temperatures (small: $-1.74\,^{\circ}\text{C}$, 95% CI [-1.74, -1.56]; medium: $-0.40\,^{\circ}\text{C}$, CI [-0.52, -0.27]; large: $0.05\,^{\circ}\text{C}$, CI [-0.09, 0.19]), with peak densities of small, medium and large flocks occurring at -3.81, 1.91, and $1.92\,^{\circ}\text{C}$, respectively (Figure 5A). Average minimum temperatures associated with the species' occurrence were considerably higher than that of background points (−3.59 °C, CI [−3.62, −3.56]).

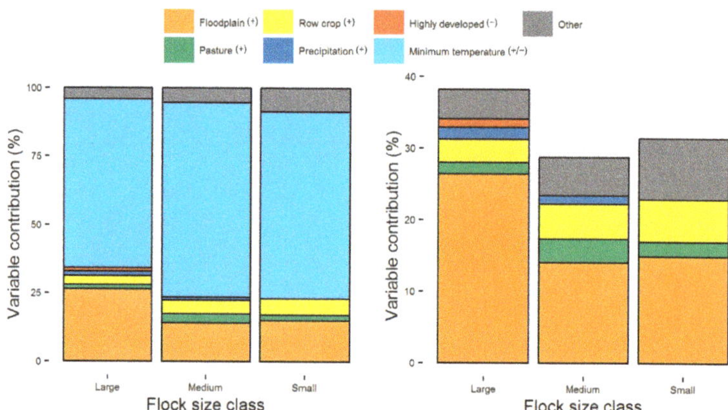

Figure 4. The relative contribution of environmental covariates to predicted Rusty Blackbird occurrence. Variables are subset to those with the greatest explanatory power, with the bar graph on the right excluding minimum temperature (i.e., showing landcover variables only). Legend symbols (+,−) represent whether Rusty Blackbirds were positively or negatively associated with the variable. Rusty Blackbird observations peaked at intermediate minimum temperatures, and is thus denoted with +/−. For a complete list of variable contributions, see Table S1.

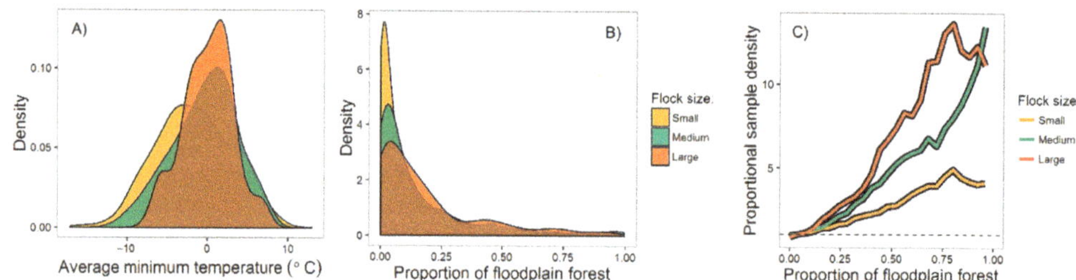

Figure 5. The gaussian kernel density distributions of average minimum temperature and floodplain forest (**A**,**B**) at occurrence points for each flock size class. Plot (**C**) describes the proportional density distribution of floodplain forest, which is the ratio of the kernel density of observations relative to that of the background points.

Among land cover covariates (Figures 4 and 5, Table 2), proportion of floodplain forest had the greatest influence on predicted habitat suitability. The variable contribution of floodplain forest to large flock distributions was nearly twice as that of small and medium flocks. Other landcover variables that contributed to predicted habitat suitability across all flock size classes were proportion of row crops and pasture, both of which were positive influences (Table 2). Proportion of highly developed land was negatively associated with large flock distributions but was not predictive of small or medium-sized flocks. Likewise, woodland and shrub land cover classes were negatively associated with small and medium-sized flocks but were not predictive of large flocks. Surprisingly, the data provided only limited support for a positive association between habitat suitability and non-floodplain wetlands, with emergent wetlands only associated with medium-sized flock distributions and wooded wetlands for only small flocks.

Table 2. Land cover variable coefficients (λ) and contribution of each variable to model performance and small, medium, and large flocks. Variables represent the proportional cover of the land use category within a 4 km grid cell. Empty cells represent variables that were excluded from the candidate model sets due to lack of statistical support, as evaluated with AIC_c.

Land Cover Variable	Variable Coefficient (λ) Flock Size			Variable Contribution (%) Flock Size		
	Small	Medium	Large	Small	Medium	Large
Highly developed			−0.83			1.2
Low-intensity development		1.17	1.41		1.3	1.1
Floodplain	1.84	2.44	3.33	14.9	14.1	26.4
Mixed forest			2.33			2.9
Pasture	0.70	1.58	1.47	2.1	3.3	1.6
Row crop	1.09	1.93	2.08	5.9	4.9	3.3
Shrub	−3.26	−2.29		3.7	1.1	
Emergent wetand		2.50			1.5	
Woody Wetland	1.63			2.9		
Woodland	−1.73	−1.64		1.9	1.3	

8. Environmental Niche

Niche breadth decreased with increasing flock size. Predicted niche breadth for small flocks was 0.67, medium flocks were 0.47, and large flocks 0.37. Prevalence (i.e., average probability of presence at background sites) was inversely associated with flock size, with small, medium, and large flocks having predicted prevalence values of, 0.44, 0.28, and 0.20, respectively. The observed prevalence for large flocks was significantly lower than that of small flocks ($p < 0.001$) but statistically indistinguishable from medium-sized flocks ($p = 0.773$). Observed prevalence for medium flocks was significantly lower than that of small flocks ($p = 0.005$). Likewise, the fractional predicted area of suitable habitat increased with flock size, ranging from 33 percent for small flocks, 24 percent for medium-sized flocks, and 20 percent for large flocks. Fractional predicted area of suitable habitat was significantly lower for large and medium-sized flocks relative to small flocks ($p < 0.001$) but was statistically indistinguishable for large and medium flocks ($p = 0.72$).

The three flock size classes occupied similar but statistically distinct realized environmental niches. Across niche axes, we found significant differences between the small flock size class and the medium ($i = 0.939$, $p < 0.001$) and large flock size classes ($i = 0.906$, $p < 0.001$) but only marginal differences between the medium and large flock size classes ($i = 0.979$, $p = 0.102$). Among individual environmental variables, we found strong evidence of differences between the distribution of samples across flock size classes, most notably regarding the distribution about minimum temperatures and floodplain forests (Table 3). Whereas the modes of the temperature distributions for all three flock size classes were similar, medium and large flocks were observed in a narrower range of temperatures than small flocks (Figure 5B). Likewise, while all flock size classes were positively associated with floodplain forest (Figure 5B), the degree of association to floodplain forest increased with flock size (Figure 5C).

Table 3. The difference (D) in the distribution of each environmental covariate between Rusty Blackbird flock size classes. *p*-values below 0.05, in bold, suggest that the flocks occupy a different environmental space.

Environmental Variable	Small vs. Medium Flocks		Small vs. Large Flocks		Medium vs. Large Flocks	
	D	*p*-Value	D	*p*-Value	D	*p*-Value
High-intensity dev.	0.058	0.197	0.084	0.164	0.066	0.573
Low-intensity dev.	0.053	0.324	0.057	0.623	0.360	0.993
Floodplain forest	**0.112**	**<0.001**	**0.208**	**<0.001**	**0.139**	**0.001**
Hardwood forest	**0.096**	**0.005**	0.045	0.866	0.113	0.055
Mixed forest	0.024	0.994	0.099	0.064	0.096	0.153
Grassland	**0.117**	**<0.001**	0.084	0.164	0.079	0.338
Pasture	0.046	0.489	0.055	0.663	0.062	0.651
Row crop	0.036	0.788	0.098	0.066	0.096	0.146
Shrub	0.028	0.956	**0.014**	**0.002**	**0.141**	**0.007**
Upland forest	0.046	0.489	0.022	1.000	0.050	0.873
Emergent wetland	0.034	0.836	0.099	0.065	0.0696	0.516
Woody wetland	0.061	0.183	**0.116**	**0.018**	0.087	0.240
Woodland	**0.010**	**0.003**	**0.139**	**0.002**	0.040	0.979
Precipitation	**0.088**	**0.013**	**0.153**	**0.001**	0.093	0.179
Minimum temperature	**0.170**	**<0.001**	**0.271**	**<0.001**	**0.112**	**0.030**

9. Discussion

We used novel targeted citizen science data from the Blitz and eBird to identify and describe suitable habitat for a widespread, highly vagile, and declining species, the Rusty Blackbird. Across two winters, occurrence was most strongly linked to minimum temperatures and the proportion of floodplain forest in the surrounding landscape. Although all flock sizes were associated with proportion of floodplain forest, pasture, and row crops, and the convex quadratic effect of average minimum temperature, we found considerable differences among flock size distributions. Predictably, large and medium-sized flocks were more similar in their environmental distributions relative to small flocks. Large and medium flocks had narrower environmental niches than small flocks, showing a greater preference for floodplain forest. We identified Rusty Blackbird hotspots in the LMAV, the Black Belt region of Alabama and Mississippi and the Southeast Coastal Plain. The Blitz was ultimately successful: adding Blitz data to eBird data significantly increased our predictive power.

10. Environmental Predictors of Occurrence

Rusty Blackbirds exhibited a strong association with floodplain forest, with the degree of influence of this variable positively associated with flock size (Table 3, Figure 5). Given that floodplain forest is expected to represent optimal foraging habitat for Rusty Blackbirds, this provides supportive evidence that large flocks are representative of higher quality habitats as a consequence of coarse-level local enhancement (e.g., [22,23]). Winter floodplains in the southeastern United States historically provided vast areas of shallow water crucial for migratory birds [61]. These environments, which represent seasonally flooded forested habitats, provide an abundance of invertebrates for Rusty Blackbirds, which spend much of the winter foraging in, or adjacent to, shallow water [1,62]. In the southeastern United States, floodplain forests have experienced extraordinary rates of direct landscape modification, historically due primarily to agricultural conversion and more recently as a result of urban development [9,61,63,64]. Moreover, hydrologic alteration has greatly reduced the extent and integrity of floodplain wetlands throughout the region [65]. Combined, these pressures have resulted in a 75–85% loss of the historic extent of floodplain forests in the Southeast [66,67] and the remnant patches of this habitat are highly fragmented [68,69]. Within our study extent, the LMAV hosts the largest area of floodplain forest and this region yielded the highest predicted habitat suitability for Rusty Blackbirds across flock size classes (Figure 2). Christmas Bird Count data suggest that the LMAV, considered to be the core of the species' range, has experienced the steepest population declines and thus, maintenance of floodplain forests in this region is likely critical for Rusty Blackbird conservation [70].

Surprisingly, neither emergent nor woody wetlands had substantial influence on habitat suitability for Rusty Blackbirds. Although these land cover classes appear to provide important habitat features for the species [71], they may have been underrepresented by Blitz and eBird samples. Our ability to detect an influence of these land cover classes may have also been limited by the coarse spatial grain of our analysis. For example, though Rusty Blackbirds often forage on wet lawns in suburban areas [25] it would not be feasible to detect these habitat features at our scale of analysis, especially when using traveling count data. Sampling constraints also limited our ability to assess the effects of pecan orchards on Rusty Blackbird occurrences. Lipid-rich tree mast from pecan orchards may help the birds prepare for cold weather events [25], and wintering individuals that utilized pecan groves in Mississippi (mostly adult males), were in better body condition than those in wet forest or along creeks [24]. Despite the expected importance of this anthropogenic resource, there was not adequate representation of this land cover class in background nor Rusty Blackbird samples to include the variable in our analysis.

The influence of human land use on habitat suitability varied with flock size. Large flocks, unlike medium or small flocks, were negatively associated with high intensity development; i.e., when the proportional cover of high intensity development exceeded 70%, the probability of encountering a large flock approached zero. These results are consistent with those of Mettke-Hofmann et al. [24], Newell [10], and Newell Wohner et al. [25] who observed that smaller groups of Rusty Blackbirds can use certain agricultural and suburban areas, taking advantage of abundant pecans and earthworms, respectively. The species' positive (though relatively weak) association with pasture and row crops is more puzzling, as it is not generally associated with these habitats (pecans are not classified as a row crop; [1]). It is possible that the low-lying, nutrient-rich floodplain landscapes that the species favors also make good croplands and pasture for farmers, and as such, both floodplain forest that Rusty Blackbirds presumably prefer, and that these agricultural habitats were often included in the 4-km grid cells we employed. Combined, our findings suggest that, though suburban and agricultural areas appear to offer resources for smaller groups of Rusty Blackbirds, larger tracts of floodplain forest will have to be maintained to support the persistence of large flocks.

Average minimum temperature was the most important variable driving habitat suitability for all flock sizes, with medium and large flocks predicted to occur at higher temperatures than small flocks. Habitat suitability estimates for medium and large flocks peaked where minimum mean temperatures occurred at means slightly higher than the freezing point of water: 1.91 °C and 1.92 °C, respectively. Suitability estimates for small flocks peaked at −3.8 °C—nearly 6 °C less than medium and large flocks. The warmer temperatures associated with medium and large flocks likely reflect the foraging behavior of Rusty Blackbirds, which spend much of their time wading in shallow water searching for aquatic prey [1]. Shallow water tends to freeze before deeper water, which would quickly prohibit birds from foraging in the substrate, making aquatic prey unavailable. Moreover, wetland invertebrate abundance is positively associated with water temperature [72–74]. Conversely, the colder temperatures associated with small flocks may represent marginal habitats in which birds utilize terrestrial resources in freezing temperatures (see [24,25]).

Because temperatures were averaged across years of the study, we were unable to assess the annual occurrences of Rusty Blackbirds. The species varies widely in its wintering distribution [9] and site fidelity [10], most likely because of differences in winter temperatures and water levels across years. Rusty Blackbirds' association with temperatures just above freezing supports our theory that the species, which is a facultative migrant not tightly linked to day length [1], migrates south from the Boreal forest to avoid freezing water. Birds may then spend many weeks in the northern United States during stopover periods in fall [75,76], only to move south again when frozen foraging habitat pushes them further. Future work should assess whether carryover effects due to variable wintering conditions impact the birds across their full annual cycle [77,78], including on their breeding grounds in the Boreal forest.

11. Larger Flocks Have a Narrower Niche Breadth

We found strong evidence that large and medium flocks have narrower environmental niches than small flocks, as measured by Levins' [29] niche breadth metric and, consequently, prevalence and fractional predicted area of suitable habitat was higher for small flocks than medium and large flock size classes. Whereas such differences in niche breadth have been observed across a variety of taxa using environmental niche modeling (e.g., [79]), this represents among precious few examples of the relationship between flock size and niche [26]. This pattern is abundantly clear across large scales, as small flocks have a much broader geographic range of suitable habitat than large and medium flocks. Conversely, the area of suitable habitat (Figure 2) for large flocks is much smaller and is largely constrained within the center of the species range. Again, this highlights the relative vulnerability of larger flocks of Rusty Blackbirds and the relative flexibility of smaller flocks. Small flocks may be able to persist in heterogenous landscapes with small patches of ephemeral resources (e.g., fruiting pecan orchards, lawns with surfacing earthworms), whereas larger flocks may be limited to the highest quality habitats. Future studies (e.g., using tracking technology) should examine space use of large vs. small flocks to understand how medium (i.e., <4 km) and small-scale habitat features affect fitness and niche space.

The use of flock size herein to approximate habitat quality represents a limitation to the inference of our results. Blitz and traditional eBird observations were traveling counts, thus in some instances, birds observed on a given count could have comprised one or several true flocks. As such, it was not technically possible to distinguish between a single large flock or a higher density of smaller flocks; however, given our own observations of the species (e.g., [12]), we believe the vast majority of the counts probably represented single flocks, so we are comfortable interpreting results based on this assumption. Moreover, traveling counts limit the spatial grain of the analysis, as counts may occur at any point along the observer's route. To address this, we highly recommend the use of stationary count data to allow for assessing the distribution of flocks as it relates to land cover data (e.g., evaluating resource dispersion hypothesis as described by [80]). Even without these limitations, density itself is not a clear indicator of habitat quality [11,81]. Further research (e.g., survival) is therefore needed to confirm that large flocks are representative of environments of higher quality habitat [33]. Sex or age-related segregation of the observed flocks may provide additional evidence of habitat quality [82]. Dominance hierarchies can indicate habitat quality in a number of bird species (e.g., the American Redstart, *Setophaga ruticilla*, [83]). For example, in Rusty Blackbirds, Mettke-Hofmann et al. [24] observed that adult males maintained a higher body condition and occupied habitats with higher pecan mast production than females and immature birds. Unfortunately, age and sex data from the Blitz were not adequate to address this issue in the current study.

Ultimately, a demographic response, most appropriately overwinter survival, is necessary to truly ascertain whether flock size is an appropriate indication of winter habitat quality for Rusty Blackbirds [84]. In particular, despite some location-specific evidence hinting at population limitation on the breeding grounds in the Boreal forest [7,8], a full-annual cycle population model suggests that wintering juvenile survival is the demographic parameter most tied to the species' rate of population change (Clark Rushing, Steve Matsuoka, Luke L. Powell et al. unpublished analysis). Whereas estimating overwinter survival may be very challenging given the species' low site fidelity, alternative proximate habitat quality metrics that can be utilized in combination for Rusty Blackbird flocks include but are not limited to: departure time (e.g., [85]), body condition (e.g., [86,87], but see [84,88]), and telomere length (e.g., [89]).

12. Regional Hotspots and Conservation

The LMAV was a particularly suitable area for all flock sizes (Figure 2, [5]). Though we also found a broad swath across the southeast coastal plain that was generally suitable, with small hotspots in north coastal South Carolina [25], the majority of the East Coast of the United States appears less suitable than the LMAV. Stable isotope data [90] and light level geolocators [75] indicate that Rusty Blackbirds wintering in the LMAV primarily migrate to Alaska and western Canada—the Western Boreal—whereas those from the southeast coastal plain breed in northeast North America—the Southeastern Boreal. It remains unclear whether flock sizes in the southeast coastal plain are constrained by anthropogenic disturbance or if it is the quality and extent of floodplain forests of the LMAV that leads to larger flock sizes in that region.

A novel finding from our study was the emergence of the Black Belt region (Ecoregion 65, [91–93]) as a hotspot across all Rusty Blackbird flock size classes. The Black Belt, which arcs along the boundary of the piedmont and coastal plain in Mississippi and Alabama, was historically a mosaic of prairie and forested habitat [94]. Whereas most of the region's native vegetation was cleared for cotton-based agriculture [94–97], the Black Belt remains embedded within a vast matrix of floodplain forest. To date, we know of no Rusty Blackbird research projects within 500 km of the Black Belt—thus, our results highlight a need for targeted field studies to further qualify the use and importance of this region to Rusty Blackbirds.

The emergence of the Black Belt region as a hotspot underscores the benefits and limitations of species distribution modeling. Whereas comparably few eBird and Blitz samples were recorded within the region, the MaxEnt model, as implemented here, is carried out in environmental space [52], which allowed us to generate predictions for the Black Belt as a function of its environmental characteristics [98]. As a cautionary note, however, this approach is limited in its ability to directly inform conservation efforts, as it yields information about the type of environment an organism uses rather than occurrences per se [99]. Moreover, our models are undoubtedly imperfect—for example, some areas of high predicted suitability for large flocks were of low predicted suitability for medium-sized flocks (e.g., north-central LMAV). This was likely driven by the difference in habitat preferences of large vs. medium flocks (e.g., large flocks had higher preference for floodplain forest, Table 3) as well as by stochasticity in where birds were detected. As such, we suggest that systematic sampling of predicted Rusty Blackbird hotspots—particularly in the Black Belt region—is a crucial step towards linking our results with management efforts.

13. Did the Blitz Help Relative to eBird Alone?

The Blitz significantly improved the predictive power of spatial models compared to using eBird alone (Figure 3), highlighting the value of this novel citizen science approach to improve understanding of a wide-ranging species of conservation concern. Further, the addition of the Blitz especially improved our predictive power for large flocks (Figure 3B), as we effectively doubled the number of large flocks detected (Table 1). This is important, as it allows us to precisely concentrate targeted studies on sites at which we predict only the very highest flock sizes of Rusty Blackbirds. We also utilized a somewhat novel way to address biases associated with modeling eBird data by using background samples from eBird lists across taxa: our background data were generated with equivalent bias to our occurrence data. Our comparison of background and occurrence data therefore provides conservative estimates of Rusty Blackbirds in environmental space. An important caveat to our Blitz effort is that participants were searching for Rusty Blackbirds in habitats that were expected to be suitable for the species–this has the potential to bias suitability predictions towards environments that are pre-determined to be suitable. A preliminary analysis of our results, however, produced estimates from eBird and Blitz data that were similar, suggesting that any such bias was nominal. In addition to the quantitative benefits of the Blitz, it was clear the Blitz raised considerable awareness for the plight of the Rusty Blackbird (e.g., Audubon magazine, eBird website, conservation partners participating in the Blitz, etc.). Given that citizen scientists submitted 678 checklists specifically for the Blitz, the program was clearly successful in engaging birders with ornithological research and conservation concerns to which they may not previously have been exposed.

Taken together, the data provided by citizen science participants of the Blitz and eBird program have enhanced our understanding of the occurrence of this Boreal-breeding species throughout its wintering areas. Whereas wintering Rusty Blackbirds can be found across most of the southeastern United States, only a narrow subset of the region was found to be suitable for large flocks of the species. By targeting these regions for future research, we can maximize sampling efficiency by searching areas predicted to provide high quality habitat. Targeted research is crucial given the limitations associated with presence-only observations. Using the MaxEnt framework, with presence-only data, locations where birds were not observed (i.e., background data) cannot be treated as true absences (see [100,101]). This is especially problematic when detection probability varies and cannot be incorporated into the model [102] and imparts a limitation to how citizen science data can be used to inform conservation in this context [103]. The use of presence-absence data to model occupancy (e.g., [2]), can also allow researchers to address temporal variation in the presence of Rusty Blackbird flocks, which may yield further insight into habitat quality (e.g., abundance-occupancy relationships, [104]). Moreover, identifying and counting individual flocks from fixed points in the landscape and evaluating these observations at multiple spatial scales would provide researchers with the ability to assess alternative explanations for flock size distributions, such as the resource dispersion hypothesis (reviewed in [80]). Future work that utilizes systematic sampling to estimate the geographic distribution of Rusty Blackbird flocks and assess the relationship between flock size and habitat quality is therefore a critical next step towards conserving this imperiled species.

Supplementary Materials: The following are available online at https://www.mdpi.com/1424-2818/13/2/62/s1: Table S1: rubl_gap_reclassification.csv, Supplementary File S1: Additional methods and results, Supplementary File S2: rubl_distribution_maps.zip.

Author Contributions: Conceptualization, R.S.G.; methodology, B.S.E.; formal analysis, B.S.E.; data curation, B.S.E.; writing—original draft preparation, B.S.E., L.L.P., D.W.D., and S.M.B.; supervision, R.S.G.; project administration, R.S.G.; funding acquisition, D.W.D. All authors have read and agreed to the published version of the manuscript.

Funding: This research and the associated APC were supported by the U.S. Fish & Wildlife Service/USFWS.

Institutional Review Board Statement: Not applicable.

Informed Consent Statement: Not applicable.

Data Availability Statement: Publicly available datasets were analyzed in this study. This data can be found here: https://github.com/SMBC-NZP/rubl.

Acknowledgments: Russ Greenberg conceived of, advertised, and led the execution of the Blitz; our hope is that this manuscript honors the memory of a brilliant ornithologist. This paper was written in association with the International Rusty Blackbird Working Group (RustyBlackbird.org; founded and led by Russ). We thank the United States Fish and Wildlife Service for funding. We deeply thank the regional coordinators and many citizen scientists that contributed to eBird and to the Blitz, and we thank the Cornell Lab of Ornithology and eBird for creating a portal for Blitz data entry. The findings and conclusions in this article are those of the author(s) and do not necessarily represent the views of the U.S. Fish and Wildlife Service. Any mention of trade names is purely coincidental and does not represent endorsement by the author(s) and / or their respective organization.

Conflicts of Interest: The authors declare no conflict of interest.

References

1. Avery, M.L. Rusty Blackbird (*Euphagus Carolinus*), version 1.0. In *Birds of the World*; Poole, A.F., Ed.; Cornell Lab of Ornithology: Ithaca, NY, USA, 2020. Available online: https://birdsoftheworld.org/bow/species/rusbla/cur/introduction (accessed on 10 February 2018).
2. Luscier, J.D.; Lehnen, S.E.; Smith, K.G. Habitat Occupancy by Rusty Blackbirds Wintering in the Lower Mississippi Alluvial Valley. *Condor* **2010**, *112*, 841–848. [CrossRef]
3. Powell, L.L.; Hodgman, T.P.; Fiske, I.J.; Glanz, W.E. Habitat occupancy of Rusty Blackbirds (Euphagus carolinus) breeding in northern New England, USA. *Condor* **2014**, *116*, 122–133. [CrossRef]
4. Greenberg, R.; Droege, S. On the Decline of the Rusty Blackbird and the Use of Ornithological Literature to Document Long-Term Population Trends. *Conserv. Biol.* **1999**, *13*, 553–559. [CrossRef]
5. Greenberg, R.; Demarest, D.W.; Matsuoka, S.M.; Mettke-Hofmann, C.; Evers, D.; Hamel, P.B.; Luscier, J.; Powell, L.L.; Shaw, D.; Avery, M.L.; et al. Understanding Declines in Rusty Blackbirds. *Boreal Birds North Am.* **2011**, *41*, 107–126. [CrossRef]
6. Sauer, J.R. *The North American Breeding Bird Survey, Results and Analysis 1966–2015. Version 1.30.2015. USGS Patuxent Wildlife Research Center: Laurel, MD 2015.* Available online: http://www.mbr-pwrc.usgs.gov.bbs/ (accessed on 15 August 2016).
7. Powell, L.L.; Hodgman, T.P.; Glanz, W.E.; Osenton, J.D.; Fisher, C.M. Nest-Site Selection and Nest Survival of the Rusty Blackbird: Does Timber Management Adjacent to Wetlands Create Ecological Traps? *Condor* **2010**, *112*, 800–809. [CrossRef]
8. Edmonds, S.T.; Evers, D.C.; Cristol, D.A.; Mettke-Hofmann, C.; Powell, L.L.; McGann, A.J.; Armiger, J.W.; Lane, O.P.; Tessler, D.F.; Newell, P.; et al. Geographic and Seasonal Variation in Mercury Exposure of the Declining Rusty Blackbird. *Condor* **2010**, *112*, 789–799. [CrossRef]
9. Greenberg, R.; Matsuoka, S.M. Special section: Rangewide ecology of the declining Rusty Blackbird. Rusty Blackbird: Mysteries of a species in decline. *Condor* **2010**, *112*, 770–777. [CrossRef]
10. Newell, P.J. Winter Ecology of the Rusty Blackbird (Euphagus Carolinus). Ph.D. Thesis, University of Georgia, Athens, GA, USA, 2013.
11. Van Horne, B. Density as a Misleading Indicator of Habitat Quality. *J. Wildl. Manag.* **1983**, *47*, 893–901. [CrossRef]
12. Borchert, S.M. Site-Specific Habitat and Landscape Associations of Rusty Blackbirds Wintering in Louisiana. Master's Thesis, Louisiana State University, Baton Rouge, LA, USA, 2015.
13. eBird Basic Dataset. *Version: EBD_relNov-2017*; Cornell Lab of Ornithology: Ithaca, NY, USA, 2017; p. 8.
14. Beauchamp, G.; Belisle, M.; Giraldeau, L.-A. Influence of conspecific attraction on the spatial distribution of learning foragers in a patchy habitat. *J. Anim. Ecology* **1997**, *66*, 671–682. [CrossRef]
15. Pöysä, H. Group Foraging in Patchy Environments: The Importance of Coarse-Level Local Enhancement. *Ornis Scand.* **1992**, *23*, 159. [CrossRef]
16. Stolen, E.D.; Collazo, J.A.; Percival, H.F. Group-Foraging Effects on Capture Rate in Wading Birds. *Condor* **2012**, *114*, 744–754. [CrossRef]
17. Thiebault, A.; Mullers, R.H.; Pistorius, P.A.; Tremblay, Y. Local enhancement in a seabird: Reaction distances and foraging consequence of predator aggregations. *Behav. Ecology* **2014**, *25*, 1302–1310. [CrossRef]
18. Ward, P.; Zahavi, A. The importance of certain assemblages of birds as "information-centres" for food-finding. *Ibis* **1973**, *115*, 517–534. [CrossRef]
19. Pride, R.E. Optimal group size and seasonal stress in ring-tailed lemurs (*Lemur catta*). *Behav. Ecol.* **2005**, *16*, 550–560. [CrossRef]
20. Hoare, D.; Couzin, I.; Godin, J.-G.; Krause, J. Context-dependent group size choice in fish. *Anim. Behav.* **2004**, *67*, 155–164. [CrossRef]
21. Caraco, T. Time Budgeting and Group Size: A Theory. *Ecology* **1979**, *60*, 611–617. [CrossRef]
22. Beerens, J.M.; Noonburg, E.G.; Gawlik, D.E. Linking dynamic habitat selection with wading bird foraging distributions across resource gradients. *PLoS ONE* **2015**, *10*, e0128182. [CrossRef]
23. Gopi Sundar, K.S. Flock Size, Density and Habitat Selection of Four Large Waterbirds Species in an Agricultural Landscape in Uttar Pradesh, India: Implications for Management. *Waterbirds* **2006**, *29*, 365–374.
24. Mettke-Hofmann, C.; Hamel, P.B.; Hofmann, G.; Zenzal, T.J., Jr.; Pellegrini, A.; Malpass, J. Competition and Habitat Quality Influence Age and Sex Distribution in Wintering Rusty Blackbirds. *PLoS ONE* **2015**, *10*, e0123775. [CrossRef]
25. Wohner, P.J.; Cooper, R.J.; Greenberg, R.S.; Schweitzer, S.H. Weather affects diet composition of rusty blackbirds wintering in suburban landscapes. *J. Wildl. Manag.* **2015**, *80*, 91–100. [CrossRef]
26. Arevalo, J.E.; Gosler, A.G. The behaviour of Treecreepers *Certhia familiaris* in mixed-species flocks in winter. *Bird Study* **1994**, *41*, 1–6. [CrossRef]
27. Colwell, R.K.; Futuyma, D.J. On the Measurement of Niche Breadth and Overlap. *Ecology* **1971**, *52*, 567–576. [CrossRef]
28. Hutchinson, G.E. Concluding remarks. *Cold Spring Harb. Symp. Quant. Biol.* **1957**, *22*, 415–427. [CrossRef]
29. Levins, R. *Evolution in Changing Environments*; Princeton University Press: Princeton, NJ, USA, 1968.
30. Sullivan, B.L.; Wood, C.L.; Iliff, M.J.; Bonney, R.E.; Fink, D.; Kelling, S. eBird: A citizen-based bird observation network in the biological sciences. *Biol. Conserv.* **2009**, *142*, 2282–2292. [CrossRef]
31. Evans, B.S. *Summary Report: Assessing Rusty Blackbird Habitat Suitability on Wintering Grounds and during Spring Migration Using A Large Citizen-Science Dataset*; Smithsonian Migratory Bird Center: Washington, DC, USA, 2016. Available online: http://rustyblackbird.org/wp-content/uploads/RUBL_Blitz10pgReportEvans23jan2017.pdf (accessed on 10 February 2018).

32. Hirzel, A.H.; Le Lay, G. Habitat suitability modelling and niche theory. *J. Appl. Ecology* **2008**, *45*, 1372–1381. [CrossRef]
33. Kellner, C.J.; Brawn, J.D.; Karr, J.R. What Is Habitat Suitability and How Should It Be Measured? In *Wildlife 2001: Populations*; McCullough, D.R., Barrett, R.H., Eds.; Elsevier Applied Science: New York, NY, USA, 1992; pp. 476–488.
34. Phillips, S.J.; Anderson, R.P.; Schapire, R.E. Maximum entropy modeling of species geographic distributions. *Ecology Model.* **2006**, *190*, 231–259. [CrossRef]
35. Sullivan, B.L.; Aycrigg, J.L.; Barry, J.H.; Bonney, R.E.; Bruns, N.; Cooper, C.B.; Damoulas, T.; Dhondt, A.A.; Dietterich, T.; Farnsworth, A.; et al. The eBird enterprise: An integrated approach to development and application of citizen science. *Biol. Conserv.* **2014**, *169*, 31–40. [CrossRef]
36. Rowcliffe, M. Activity: Animal Activity Statistics. R Package Version 1.3. 2019. Available online: https://CRAN.R-project.org/package=activity (accessed on 15 July 2019).
37. Nouvellet, P.; Rasmussen, G.S.A.; Macdonald, D.W.; Courchamp, F. Noisy clocks and silent sunrises: Measurement methods of daily activity pattern. *J. Zool.* **2011**, *286*, 179–184. [CrossRef]
38. Vazquez, C.; Rowcliffe, J.M.; Spoelstra, K.; Jansen, P.A. Comparing diel activity patterns of wildlife across latitudes and seasons: Time transformations using day length. *Methods Ecology Evol.* **2019**, *10*, 2057–2066. [CrossRef]
39. Ridout, M.S.; Linkie, M. Estimating overlap of daily activity patterns from camera trap data. *J. Agric. Biol. Env. Stat.* **2009**, *14*, 322–337. [CrossRef]
40. Dickinson, J.L.; Zuckerberg, B.; Bonter, D.N. Citizen science as an ecological research tool: Challenges and benefits. *Annu. Rev. Ecology Evol. Syst.* **2010**, *41*, 149–172. [CrossRef]
41. Hortal, J.; Jiménez-Valverde, A.; Gómez, J.F.; Lobo, J.M.; Baselga, A. Historical bias in biodiversity inventories affects the observed environmental niche of the species. *Oikos* **2008**, *117*, 847–858. [CrossRef]
42. Newbold, T.; Reader, T.; El-Gabbas, A.; Berg, W.; Shohdi, W.M.; Zalat, S.; El Din, S.B.; Gilbert, F. Testing the accuracy of species distribution models using species records from a new field survey. *Oikos* **2010**, *119*, 1326–1334. [CrossRef]
43. Dudík, M.; Phillips, S.J.; Schapire, R.E. Correcting sample selection bias in maximum entropy density estimation. *Adv. Neural Inf. Process. Syst.* **2006**, *18*, 323–330.
44. Elith, J.; Phillips, S.J.; Hastie, T.; Dudík, M.; Chee, Y.E.; Yates, C.J. A statistical explanation of MaxEnt for ecologists. *Divers. Distrib.* **2010**, *17*, 43–57. [CrossRef]
45. Phillips, S.J.; Dudík, M.; Elith, J.; Graham, C.H.; Lehmann, A.; Leathwick, J.; Ferrier, S. Sample selection bias and occurrence-only distribution models: Implications for background and pseudo-absence data. *Ecology Appl.* **2009**, *19*, 181–197. [CrossRef]
46. Anderson, R.P.; Raza, A. The effect of the extent of the study region on GIS models of species geographic distributions and estimates of niche evolution: Preliminary tests with montane rodents (genus Nephelomys) in Venezuela. *J. Biogeogr.* **2010**, *37*, 1378–1393. [CrossRef]
47. Hijmans, R.J.; van Etten, J.; Cheng, J.; Mattiuzzi, M.; Sumner, M.; Greenberg, J.A.; Lamigueiro, O.P.; Bevan, A.; Racine, E.B.; Shortridge, A.; et al. Raster: Geographic Data Analysis and Modeling. In *R Package Version 2.8-19*. 2015. Available online: https://CRAN.R-project.org/package=raster (accessed on 27 September 2020).
48. R Core Team 2013. *R: A Language and Environment for Statistical Computing*; R Foundation for Statistical Computing: Vienna, Austria, 2013. Available online: http://www.r-project.org/ (accessed on 10 March 2012).
49. PRISM Climate Group. Oregon State University. Available online: http://prism.oregonstate.edu (accessed on 1 June 2013).
50. Scott, J.M.; Davis, F.; Csuti, B.; Noss, R.; Butterfield, B.; Groves, C.; Anderson, H.; Caicco, S.; D'Erchia, F.; Edwards, T.C., Jr.; et al. Gap Analysis: A Geographic Approach to Protection of Biological Diversity. In *Wildlife Monographs*; John Wiley & Sons: Hoboken, NJ, USA, 1993; pp. 3–41.
51. Hijmans, R.J.; Phillips, S.; Leathwick, J.; Elith, J.; Hijmans, M.R.J. Package 'Dismo'. *Circles* **2017**, *9*, 1–68.
52. Merow, C.; Smith, M.J.; Silander, J.A. A practical guide to MaxEnt for modeling species' distributions: What it does, and why inputs and settings matter. *Ecography* **2013**, *36*, 1058–1069. [CrossRef]
53. Duan, R.Y.; Kong, X.Q.; Huang, M.Y.; Fan, W.Y.; Wang, Z.G. The predictive performance and stability of six species distribution models. *PLoS ONE* **2014**, *9*, e112764. [CrossRef]
54. Swets, J.A. Measuring the accuracy of diagnostic systems. *Science* **1988**, *240*, 1285–1293. [CrossRef]
55. Shcheglovitova, M.; Anderson, R.P. Estimating optimal complexity for ecological niche models: A jackknife approach for species with small sample sizes. *Ecology Model.* **2013**, *269*, 9–17. [CrossRef]
56. Simpson, G.L. Permute: Functions for Generating Restricted Permutations of Data. R Package Version 0.9-5. 2019. Available online: https://CRAN.R-project.org/package=permute (accessed on 10 March 2012).
57. Warren, D.L.; Glor, R.E.; Turelli, M. ENMTools: A toolbox for comparative studies of environmental niche models. *Ecography* **2010**, *33*, 607–611. [CrossRef]
58. Phillips, S.J.; Elith, J. On estimating probability of presence from use-availability or presence-background data. *Ecology* **2013**, *94*, 1409–1419. [CrossRef] [PubMed]
59. Warren, D.L.; Glor, R.E.; Turelli, M. Environmental niche equivalency versus conservatism: Quantitative approaches to niche evolution. *Evolution* **2008**, *62*, 2868–2883. [CrossRef] [PubMed]
60. Heibl, C.; Calenge, C. Phyloclim: Integrating Phylogenetics and Climatic Niche Modeling. In *R Package Version 0.9-4*. 2013. Available online: https://CRAN.R-project.org/package=phyloclim (accessed on 27 September 2020).

61. Brown, C.R.; Baxter, C.; Pashley, D.N. The Ecological Basis for the Conservation of Migratory Birds in the Mississippi Alluvial Valley. In *Strategies for Bird Conservation: The Partners in Flight Planning Process*; Bonney, R., Pashley, D.N., Cooper, R.J., Niles, L., Eds.; Cornell Lab of Ornithology: Ithaca, NY, USA, 1999; pp. 3–6.

62. Wehrle, B.W.; Kaminski, R.M.; Leopold, B.D.; Smith, W.P. Aquatic invertebrate resources in Mississippi forested wetlands during winter. *Wildlife Society Bulletin* **1995**, *23*, 774–783.

63. Faulkner, S.P. Urbanization impacts on the structure and function of forested wetlands. *Urban Ecosyst.* **2004**, *7*, 89–106. [CrossRef]

64. Lockaby, B.G. Floodplain ecosystems of the Southeast: Linkages between forests and people. *Wetl* **2009**, *29*, 407–412. [CrossRef]

65. Harmar, O.P.; Clifford, N.J.; Thorne, C.R.; Biedenharn, D.S. Morphological changes of the Lower Mississippi River: Geomorphological response to engineering intervention. *River Res. Appl.* **2005**, *21*, 1107–1131. [CrossRef]

66. Hefner, J.M.; Brown, J.P. Wetland trends in the southeastern U.S. *Wetlands* **1984**, *4*, 1–11. [CrossRef]

67. Hefner, J.M.; Wilen, B.O.; Dahl, T.E.; Frayer, W.E. *Southeastern Wetlands: Status and Trends, Mid-1970s to Mid-1980s*; U.S. Fish and Wildlife Service and U.S. Environmental Protection Agency: Atlanta, GA, USA, 1994.

68. Rudis, V.A. Regional forest fragmentation effects on bottomland hardwood community types and resource values. *Landsc. Ecology* **1995**, *10*, 291–307. [CrossRef]

69. Twedt, D.J.; Loesch, C.R. Forest area and distribution in the Mississippi alluvial valley: Implications for breeding bird conservation. *J. Biogeogr.* **1999**, *26*, 1215–1224. [CrossRef]

70. Niven, D.K.; Sauer, J.R.; Butcher, G.S. Christmas bird count provides insights into population change in land birds that breed in the boreal forest. *Am. Birds* **2004**, *58*, 10–20.

71. DeLeon, E.E. Ecology of Rusty Blackbirds Wintering in Louisiana: Seasonal Trends, Flock Compositions and Habitat Associations. Master's Thesis, Louisiana State University, Baton Rouge, LA, USA, 2012.

72. Batema, D.L.; Kaminski, R.M.; Magee, P.A. Wetland invertebrate communities and management of hardwood bottomlands in the Mississippi Alluvial Valley. In *Ecology and Management of bottomland Hardwood Systems: The State of our Understanding*; Fredrickson, L.H., King, S.L., Kaminski, R.M., Eds.; Gaylord Memorial Laboratory Special Publication 10, University of Missouri-Columbia: Puxico, MO, USA, 2005; pp. 173–190.

73. Fredrickson, L.H.; Batema, D.L. *Greentree Reservoir Management Handbook*; Gaylord Memorial Lab, University of Missouri: Puxico, MO, USA, 1992.

74. White, D.C. Lowland hardwood wetland invertebrate community and production in Missouri. *Arch. Hydrobiol.* **1985**, *103*, 509–533.

75. Johnson, J.A.; Matsuoka, S.M.; Tessler, D.F.; Greenberg, R.; Fox, J.W. Identifying Migratory Pathways Used by Rusty Blackbirds Breeding in Southcentral Alaska. *Wilson J. Ornithol.* **2012**, *124*, 698–703. [CrossRef]

76. Wright, J.R.; Powell, L.L.; Tonra, C.M. Automated telemetry reveals staging behavior in a declining migratory passerine. *Auk* **2018**, *135*, 461–476. [CrossRef]

77. Harrison, X.A.; Blount, J.D.; Inger, R.; Norris, D.R.; Bearhop, S. Carry-over effects as drivers of fitness differences in animals. *J. Anim. Ecology* **2010**, *80*, 4–18. [CrossRef]

78. Marra, P.P.; Cohen, E.B.; Loss, S.R.; Rutter, J.E.; Tonra, C.M. A call for full annual cycle research in animal ecology. *Biol. Lett.* **2015**, *11*, 20150552. [CrossRef]

79. Saupe, E.E.; Qiao, J.; Hendricks, J.R.; Portell, R.W.; Junter, S.J.; Sobseron, J.; Lieberman, B.S. Niche breadth and geographic range size as determinants of species survival on geological time scales. *Glob. Ecology Biogeogr.* **2015**, *24*, 1159–1169. [CrossRef]

80. Johnson, D.D.; Kays, R.; Blackwell, P.G.; Macdonald, D.W. Does the resource dispersion hypothesis explain group living? *Trends Ecology Evol.* **2002**, *17*, 563–570. [CrossRef]

81. Battin, J. When Good Animals Love Bad Habitats: Ecological Traps and the Conservation of Animal Populations. *Conserv. Biol.* **2004**, *18*, 1482–1491. [CrossRef]

82. Kawecki, T.J. Adaptation to Marginal Habitats. *Annu. Rev. Ecology Evol. Syst.* **2008**, *39*, 321–342. [CrossRef]

83. Powell, L.L.; Ames, E.M.; Wright, J.R.; Matthiopoulos, J.; Marra, P.P. Interspecific competition between resident and wintering birds: Experimental evidence and consequences of coexistence. *Ecology* **2020**, in press. [CrossRef]

84. Johnson, M.D. Measuring habitat quality: A review. *Condor* **2007**, *109*, 489–504. [CrossRef]

85. Studds, C.E.; Marra, P.P. Rainfall-induced changes in food availability modify the spring departure programme of a migratory bird. *Proc. R. Soc. B Biol. Sci.* **2011**, *278*, 3437–3443. [CrossRef]

86. Bearhop, S.; Hilton, G.M.; Votier, S.C.; Waldron, S. Stable isotope ratios indicate that body condition in migrating passerines is influenced by winter habitat. *Proc. R. Soc. Lond. Ser. B* **2004**, *271*, 215–218. [CrossRef]

87. Latta, S.C.; Faaborg, J. Demographic and population responses of Cape May Warblers wintering in multiple habitats. *Ecology* **2002**, *83*, 2502–2515. [CrossRef]

88. Rintamäki, P.T.; Stone, J.R.; Lundberg, A.; Moore, F. Seasonal and diurnal body-mass fluctuations for two nonhoarding species of *Parus* in Sweden modeled using path analysis. *Auk* **2003**, *120*, 658–668. [CrossRef]

89. Young, R.C.; Kitaysky, A.S.; Barger, C.P.; Dorresteijn, I.; Ito, M.; Watanuki, Y. Telomere length is a strong predictor of foraging behavior in a long-lived seabird. *Ecosphere* **2015**, *6*, 1–26. [CrossRef]

90. Hobson, K.A.; Greenberg, R.; Van Wilgenburg, S.L.; Mettke-Hofmann, C. Migratory Connectivity in the Rusty Blackbird: Isotopic Evidence From Feathers of Historical and Contemporary Specimens. *Condor* **2010**, *112*, 778–788. [CrossRef]

91. Omernik, J.M. Ecoregions: A spatial framework for environmental management. In *Biological Assessment and Criteria: Tools for Water Resource Planning and Decision Making*; Davis, W.S., Simon, T.P., Eds.; Lewis Publishers: Boca Raton, FL, USA, 1995; pp. 49–62.
92. Omernik, J.M. Perspectives on the Nature and Definition of Ecological Regions. *Environ. Manag.* **2004**, *34*, S27–S38. [CrossRef]
93. Omernik, J.M.; Griffith, G.E. Ecoregions of the Conterminous United States: Evolution of a Hierarchical Spatial Framework. *Env. Manag.* **2014**, *54*, 1249–1266. [CrossRef]
94. Webster, G.R.; Samson, S.A. On Defining the Alabama Black Belt: Historical Changes and Variations. *Southeast. Geogr.* **1992**, *32*, 163–172. [CrossRef]
95. Barone, J.A. Historical occurrence and distribution of prairies in the Black Belt of Mississippi and Alabama. *Castanea* **2005**, *70*, 170–183. [CrossRef]
96. Cleland, H.F. Black Belt of Alabama. *Geogr. Rev.* **1920**, *10*, 375–387. [CrossRef]
97. Wilson, T.H. Natural history of the black belt prairie. *J. Ala. Acad. Sci.* **1981**, *52*, 10–19.
98. Phillips, S.J.; Dudík, M.; Schapire, R.E. A Maximum Entropy Approach to Species Distribution Modeling. In Proceedings of the Twenty-First International Conference on Machine Learning, Banff, AB, Canada, 4–8 July 2004; ACM: New York, NY, USA, 2004.
99. Elith, J.; Graham, C.H.; Anderson, R.P.; Dud'ik, M.; Ferrier, S.; Guisan, A.; Hijmans, R.J.; Huettmann, F.; Leathwick, J.R.; Lehmann, A. Novel methods improve prediction of species' distributions from occurrence data. *Ecography* **2006**, *29*, 129–151. [CrossRef]
100. Guillera-Arroita, G.; Lahoz-Monfort, J.J.; Elith, J. Maxent is not a presence–absence method: A comment on Thibaud et al. *Methods Ecology Evol.* **2014**, *11*, 1192–1197. [CrossRef]
101. Li, W.; Guo, Q.; Elkan, C. Can we model the probability of presence of species without absence data? *Ecography* **2011**, *34*, 1096–1105. [CrossRef]
102. Lahoz-Monfort, J.J.; Guillera-Arroita, G.; Wintle, B.A. Imperfect detection impacts the performance of species distribution models. *Glob. Ecology Biogeogr.* **2014**, *23*, 504–515. [CrossRef]
103. Guillera-Arroita, G.; Lahoz-Monfort, J.J.; Elith, J.; Gordon, A.; Kujala, H.; Lentini, P.E.; McCarthy, M.A.; Tingley, R.; Wintle, B.A. Is my species distribution model fit for purpose? Matching data and models to applications. *Glob. Ecology Biogeogr.* **2015**, *24*, 276–292. [CrossRef]
104. Gaston, K.J.; Blackburn, T.M.; Greenwood, J.J.D.; Gregory, R.D.; Quinn, R.M.; Lawton, J.H. Abundance-occupancy relationships. *J. Appl. Ecology* **2000**, *37*, 39–59. [CrossRef]

MDPI

St. Alban-Anlage 66

4052 Basel

Switzerland

Tel. +41 61 683 77 34

Fax +41 61 302 89 18

www.mdpi.com

Diversity Editorial Office

E-mail: diversity@mdpi.com

www.mdpi.com/journal/diversity